AF411623

Adhesion Protein Protocols

METHODS IN MOLECULAR BIOLOGY™

John M. Walker, SERIES EDITOR

METHODS IN MOLECULAR BIOLOGY™

Adhesion Protein Protocols

Second Edition

Edited by

Amanda S. Coutts

Medical Sciences Division
University of Oxford
Oxford, UK

HUMANA PRESS ✳ TOTOWA, NEW JERSEY

Preface

The second edition of *Adhesion Protein Protocols* combines traditional techniques with cutting-edge and novel techniques that can be easily adapted to different molecules and cell types. These protocols are suited for both novice and expert scientists and will hopefully be used to gain further insight into the complex and incompletely understood processes that are involved in cellular adhesion. The book begins with two chapters covering novel techniques for studying cell–cell adhesion, which traditionally has been less straightforward to examine. The first chapter by Nelson et al. describes their novel technique for studying cell–cell adhesion using bowtie-shaped microwells. This technique not only allows one to control the degree to which the cells spread, but also allows manipulation of cell–substratum interactions. In the second chapter, Vogelmann and Nelson describe the use of differential centrifugation to isolate cell–cell adhesion complexes as a useful starting point to further examine adhesion protein interactions.

Chapter 3 by Nuzzi and co-workers describes the analysis of neutrophil chemotaxis, and includes methods for retroviral infection of these difficult-to-transfect cell types and time-lapse video microscopy. In Chapter 4, McGettrick and colleagues describe several in vitro assays used to study leukocyte migration through monolayers of cultured endothelial cells. Cell motility requires the formation of pseudopodia, and in Chapter 5, Wang and Klemke describe a novel technique to purify pseudopodia from migratory cells. This technique allows the physical separation of the pseudopodia from the cell body, which allows a detailed analysis of the components of pseudopodia in migratory cells while providing a novel method to study the signaling processes involved in cell migration.

The next few chapters deal with the study of cell–matrix interaction. Petroll deals with the study of cell–matrix interactions in three-dimensional culture, using the experimental model system he has developed whereby changes in focal adhesion reorganisation can be correlated with mechanical deformation of a collagen matrix. Gallant and García describe a comprehensive set of methods to quantify cellular adhesion strength. The first of these can be used to determine initial as well as long-term adhesion strength; furthermore, two biochemical assays are described that can be used to quantitate proteins involved in adhesion.

In Chapter 8, Griffith details the use of RNA interference, now widespread in its use, to study the effects of knocking down an adhesion protein, and in

Chapter 9, Carragher provides a detailed description of how to monitor the activity of calpain, an important protease that regulates cell adhesion and signaling.

Both microarray and proteomics techniques are becoming increasingly used in the laboratory and Chapters 10 (Dalby and Yarwood) and 11 (Boyd and colleagues) describe the application of these techniques to study adhesion proteins. Bioinformatics is in constant use in the laboratory but can often be overlooked as a general laboratory protocol. Adams and Engel have provided a detailed and interesting overview of how to embark on bioinformatic analysis using adhesion proteins as examples. They cover all the major methods, such as the identification of sequence homologies, examination of sequence relationships and domain organization, as well as how to use comparative genomics.

The final chapters deal with three important techniques used in the study of adhesion molecules. In Chapter 13, Wehrle-Haller provides an excellent review of the use of fluorescence recovery after photobleaching. Beningo and Wang have designed a novel system to study cells in three-dimensional culture and in Chapter 14 describe how to employ this simple system using hydrogel coated with defined matrix proteins as substrates. In the final chapter, Zuchero provides an overview of actin purification as well as how to perform in vitro actin assembly assays, two techniques commonly employed to investigate the roles of actin-binding proteins in the kinetics and morphology of actin assembly.

I would also like to thank all the authors for their contributions, John Walker for his expert advice and assistance, and Humana Press for all their efforts.

Amanda S. Coutts

Contents

Contributors

JOSEPHINE C. ADAMS • *Department of Cell Biology, Lerner Research Institute, Cleveland Clinic Foundation, Cleveland, OH*

KAREN A. BENINGO • *Department of Biological Sciences, Wayne State University, Detroit, MI*

ROBERT S. BOYD • *Proteomics Facility, MRC Toxicology Unit, University of Leicester, Leicester, UK*

LYNN M. BUTLER • *Division of Medical Sciences, Department of Physiology, The Medical School, The University of Birmingham, Birmingham, UK*

KELVIN CAIN • *Proteomics Facility, MRC Toxicology Unit, University of Leicester, Leicester, UK*

NEIL O. CARRAGHER • *Cell Adhesion-Linked Kinase Laboratory, Beatson Institute for Cancer Research, Bearsden, Glasgow, UK*

CHRISTOPHER S. CHEN • *Department of Bioengineering, University of Pennsylvania, Philadelphia, PA*

AMANDA S. COUTTS • *Medical Sciences Division, University of Oxford, Oxford, UK*

MATTHEW J. DALBY • *Centre for Cell Engineering, Institute of Biomedical and Life Sciences, University of Glasgow, Glasgow, UK*

MARTIN J. S. DYER • *Professor of Haemato-Oncology and Honorary Consultant Physician, MRC Toxicology Unit, Leicester University, Leicester, UK*

JUERGEN ENGEL • *Department of Biophysical Chemistry, Biozentrum, University of Basel, Switzerland*

NATHAN D. GALLANT • *Woodruff School of Mechanical Engineering, Petit Institute for Bioengineering and Bioscience, Georgia Institute of Technology, Atlanta, GA*

ANDRÉS J. GARCÍA • *Woodruff School of Mechanical Engineering, Petit Institute for Bioengineering and Bioscience, Georgia Institute of Technology, Atlanta, GA*

ELEN GRIFFITH • *MRC Human Genetics Unit, Western General Hospital, Edinburgh, UK*

ANNA HUTTENLOCHER • *Departments of Pediatrics and Pharmacology, University of Wisconsin Medical School, Madison, WI*

RICHARD L. KLEMKE • *Department of Pathology and Moores Cancer Center, University of California, San Diego, La Jolla, CA*

WENDY F. LIU • *Department of Bioengineering, University of Pennsylvania, Philadelphia, PA*

MARY A. LOKUTA • *Department of Pediatrics, University of Wisconsin Medical School, Madison, WI*

HELEN M. MCGETTRICK • *Division of Medical Sciences, Department of Physiology, The Medical School, The University of Birmingham, Birmingham, UK*

GERARD B. NASH • *Division of Medical Sciences, Department of Physiology, The Medical School, The University of Birmingham, Birmingham, UK*

CELESTE M. NELSON • *Life Sciences Division, Lawrence Berkeley National Laboratory, Berkeley, CA*

W. JAMES NELSON • *Department of Molecular and Cellular Physiology, Beckman Center for Molecular and Genetic Medicine, Stanford University School of Medicine, Stanford, CA*

PAUL A. NUZZI • *Department of Molecular and Cellular Pharmacology, University of Wisconsin Medical School, Madison, WI*

W. MATTHEW PETROLL • *Department of Ophthalmology, University of Texas Southwestern Medical Center, Dallas, TX*

ROGER VOGELMANN • *Department of Molecular and Cellular Physiology, Beckman Center for Molecular and Genetic Medicine, Stanford University School of Medicine, Stanford, CA*

YINGCHUN WANG • *Department of Pathology and Moores Cancer Center, University of California, San Diego, La Jolla, CA*

YU-LI WANG • *Department of Physiology, University of Massachusetts Medical School, Worcester, MA*

BERNHARD WEHRLE-HALLER • *Department of Cellular Physiology and Metabolism, Centre Medical Universitaire, University of Geneva, Geneva, Switzerland*

STEPHEN J. YARWOOD • *Molecular Pharmacology Group, Division of Biochemistry and Molecular Biology, Institute of Biomedical and Life Sciences, University of Glasgow, Glasgow, UK*

J. BRADLEY ZUCHERO • *Department of Cellular and Molecular Pharmacology, University of California, San Francisco, CA*

1

Manipulation of Cell–Cell Adhesion Using Bowtie-Shaped Microwells

Celeste M. Nelson, Wendy F. Liu, and Christopher S. Chen

Summary

Traditional methods to study cell–cell adhesion have been limited by their inability to manipulate cell–cell interactions without simultaneously affecting other microenvironmental factors. Here we describe a novel method that enables the culture of cells with precise simultaneous control of both cell–cell and cell–substratum adhesion. Using microfabricated stamps of poly(dimethylsiloxane), we construct bowtie-shaped agarose microwells into which cells can be cultured. The degree to which cells spread is controlled by the size of the microwell; cell–cell contacts form between neighboring cells within the microwell. This chapter describes the details of stamp fabrication, agarose microwell construction, and cell culture in micropatterned substrata.

Key Words: Cell–cell interaction; cadherin; microfabrication.

1. Introduction

Micropatterning has emerged as a valuable family of techniques to precisely control the cellular adhesive environment *(1)*. Adherent cells monitor and respond to their surroundings using receptors on their surfaces that bind to ligands both in the extracellular matrix (ECM) and on neighboring cells. Micropatterning allows the investigator to present these cues in spatially organized arrangements to cells in culture with micrometer precision, providing a means to explore structure–function relationships between cells and their extracellular space. Biological studies using micropatterning have examined the roles of cell–ECM *(2,3)* and cell–cell *(4–6)* interactions in the control of cellular function with a specificity that could not have been accomplished using traditional techniques.

Cells adhere to the surrounding ECM and neighboring cells through several classes of transmembrane proteins that mechanically link to the intracellular

From: *Methods in Molecular Biology, vol. 370:*
Adhesion Protein Protocols, Second Edition
Edited by: A. S. Coutts © Humana Press Inc., Totowa, NJ

cytoskeleton. Cell–ECM and cell–cell binding interactions play an important role in regulating a number of processes, including proliferation, differentiation, and gene expression. Cell adhesion to the ECM involves binding and clustering of integrins and cell spreading against the matrix. Both integrin binding and changes in cell shape appear to be important mediators of cell signaling and function *(7)*. While many techniques have been developed to manipulate cell–ECM interactions *(8,9)*, controlling cell–cell interactions is less straightforward. Early studies examining the effects of cell–cell interactions on cellular function were limited to observations of cells grown in typical culture conditions. Cells cultured at a low density in Petri dishes form few cell–cell adhesions, whereas cells that grow to a high density form many more cell–cell adhesions; observed differences between the behaviors of cells at low and high density were then correlated to the effects of cell–cell adhesion *(10,11)*. In other studies, the aggregation of cells in suspension was used to determine relative cell–cell interaction strength *(12)*. One confounding problem with the interpretation of these experiments is that the density or aggregation context also modulates other environmental cues. For example, the concentration of secreted soluble factors in the media may be greater for high-density cultures. Adhesive factors or cell spreading against the surface also varies: typically, cells at a low density are spread to a greater extent than cells at a higher density. Because cadherins, the primary adhesive molecules involved in cell–cell adhesion, are dependent on extracellular calcium, engagement of cadherin contacts can be induced by removal and then addition of extracellular calcium into the media—also known as a "calcium switch" assay *(13)*. Again, however, decreasing calcium concentration causes cells to round up and alters cell–ECM adhesion. As many investigations have shown, cell–ECM adhesion alone directly affects cell function, perhaps through biochemical, cytoskeletal, or mechanical changes *(2,8)*. An alternative method used to study cell–cell contacts applies isolated plasma membranes to cells in culture, which was observed to mimic the effects of high cell density *(14)*. After the identification of specific families of cell–cell adhesion receptors *(15)*, some groups have more recently begun to examine the effects of cadherin engagement using protein-coated beads or surfaces *(16,17)*. Whereas studies have shown that contact to a cadherin-coated bead can induce biological responses similar to cadherin-mediated contact to neighboring cells *(18)*, the physiological relevance and biological significance of engaging a cadherin molecule attached to a bead or surface are uncertain.

Here we describe a novel microfabrication method used to isolate the effects of cell–cell contact while holding constant cell–ECM adhesion by controlling the degree to which cells spread against the underlying substratum. Cells are seeded into bowtie-shaped microwells so that at most one pair of cells is cultured within each microwell, and thus each cell only contacts a single neighboring cell.

The degree of cell spreading is determined by the area of the microwell. This technique allows the control of cell–cell contacts without changing the effective culture density, the aggregation context, or the extracellular ion concentration. Furthermore, the cadherin contacts are formed with neighboring cells and are therefore physiologically relevant. Studies using these microwells have demonstrated that some effects traditionally attributed to cell–cell contact are actually caused by contact-mediated changes in cell–ECM adhesion and that the regularity of cell–cell adhesions formed in this assay can amplify the specific effects of cell–cell contact that might otherwise be lost in the heterogeneity of typical culture conditions *(5,6)*. Here we describe the procedure used to produce bowtie-shaped microwells. We describe the photolithographic methods used to generate the master templates from which rubber stamps are cast. The stamps are then used to produce agarose substrata on glass, onto which cells can be cultured and studied.

2. Materials

1. Silicon wafers (2- to 3-in. diameter, test grade, 13–17 mils, <100> orientation, 1–10 Ωcm; Silicon Sense, Nashua, NH).
2. SU-8-negative photoresist (MicroChem Corp., Newton, MA).
3. Spin-coater (Headway Research, Inc., Garland, TX).
4. Mask aligner (Karl Suss, Waterbury Center, VT).
5. Propyleneglycol methyl ether acetate developer (Sigma-Aldrich, St. Louis, MO).
6. Digital hotplate (Barnstead Thermolyne, Dubuque, IA).
7. Vacuum dessicator (Fisher Scientific, Hampton, NH).
8. (Tridecafluoro-1,1,2,2,-tetrahydrooctyl)-1-trichlorosilane (United Chemical Technologies, Bristol, PA).
9. Glass slides (Fisher Scientific).
10. Poly(dimethylsiloxane) (PDMS; Sylgard 184; Ellsworth Adhesives, Germantown, WI).
11. UVO cleaner (Jelight Corporation, Irvine, CA).
12. Superfrost glass slides (Fisher Scientific).
13. Agarose (ISCBioExpress, Kaysville, UT).
14. Ethanol (200 proof).
15. Fibronectin (Invitrogen, Carlsbad, CA).
16. Phosphate-buffered saline (PBS).

3. Methods

Our goal was to develop a technique to control cell–cell adhesion while simultaneously controlling cell spreading. Our design is a substratum consisting of bowtie-shaped islands of matrix surrounded by nonadhesive regions. Each cell of a pair adheres and spreads to fill one half of the bowtie-shaped region; the cells contact each other at the central constriction of the bowtie. We initially screened

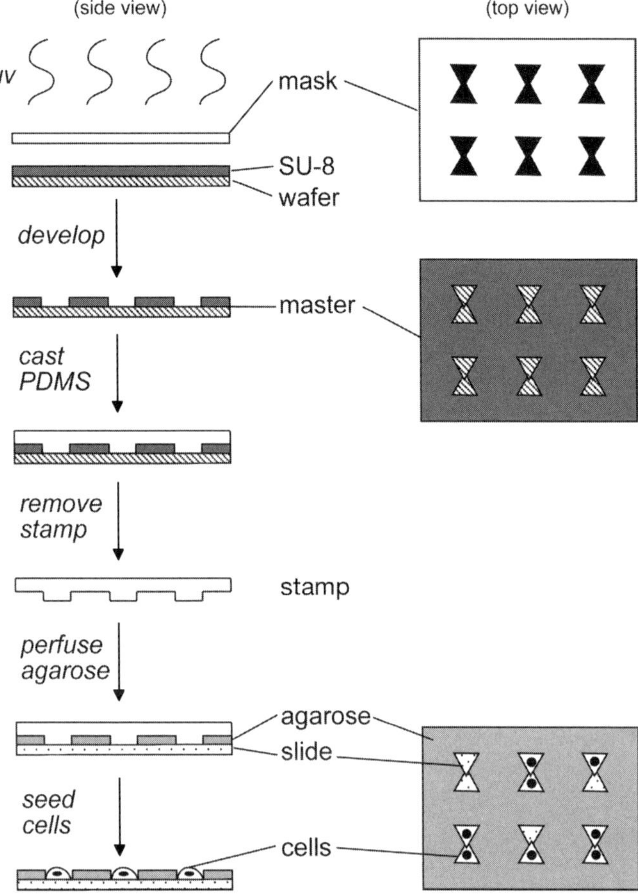

Fig. 1. Schematic of patterning process. PDMS, poly(dimethyl siloxane).

several shapes, including overlapping circles and squares, and found that endothelial cells spread best to fill the triangular features comprising one half of the
bowtie-shaped patterns. We have not tested all geometries, and other cell types
may spread preferentially in other shapes. We found that in order to precisely
control the area of cell–cell contact and prevent the contact from expanding,
we had to culture cells within shallow wells of nonadhesive material; molding
a thin film of agarose gel worked well for our purposes. The methods described
here outline the following (**Fig. 1**):

1. Fabrication of the master.
2. Molding and treatment of the stamp.
3. Construction of the agarose wells.
4. Coating of the wells and plating of cells.

3.1. Master Fabrication

1. Spin SU-8 photoresist on the shiny side of a clean silicon wafer for the specified rate and duration, depending on the desired height of microwells (a table with spin parameters for various desired thicknesses of resist can be found on the Micro-Chem website: http://www.microchem.com/products/su_eight.htm). There are many kinds of photoresist, but they all fall into two classes: positive and negative. For positive resists, the region to be removed is exposed to UV light, which alters the chemical structure of the resist and thereby renders it soluble to a particular developing solvent. The exposed resist is washed away during the developing step (*see* **step 5**), leaving a pattern of raised features on the wafer identical to the dark regions on the photomask. For negative resists, the region to be removed is blocked from exposure to UV light, which polymerizes the resist and thereby renders it insoluble to the developing solvent. The exposed resist remains on the wafer during the developing step, with the unexposed regions washing away, leaving a pattern of raised features identical to the clear regions on the photomask. Therefore, the choice of resist type (positive or negative) and design of photomask (dark-field or clear-field) go hand-in-hand. SU-8 is a negative photoresist, therefore the dark regions (bowtie shapes) on the photomask used will correspond to the eventual agarose microwells. The height of the microwells is determined by the thickness of the resist spun onto the wafer; negative resists in general, and SU-8 in particular, are suitable for making thick layers. We have found that resist thicknesses of 10–50 µm are needed to construct the agarose microwells.

2. Place the wafer on a digital hot plate for "soft bake" to evaporate solvent and dry the resist onto the surface of the wafer. The duration and temperature of baking depend on the thickness of the SU-8; a table of these parameters can also be found on the MicroChem website.

3. Place the wafer and chrome mask in the mask aligner, chrome-side down, and expose with approx 10–12 mJ/cm^2 of UV irradiation per µm thickness of resist (e.g., if making 50-µm-thick master, use 600 mJ/cm^2).

4. Remove the mask and place the wafer on a digital hot plate set at 90°C for at least 5 min ("hard bake").

5. Immerse the master into a bath of propyleneglycol methyl ether acetate developing solution and shake gently until the pattern can be observed (approx 2 min, but time will depend on thickness of the resist).

6. Dry the master under a steady stream of compressed nitrogen gas.

7. To aid in the later removal of PDMS, place the silicon master in a vacuum dessicator with a glass slide containing a drop of (tridecafluoro-1,1,2,2,-tetrahydrooctyl)-1-trichlorosilane. Evacuate the chamber. Sufficient silanization can be achieved either by exposing to vacuum overnight or by isolating the chamber for 2 h after evacuation.

3.2. Molding and Treatment of the Stamp

1. Prepare 30 g 10:1 (w/w) PDMS polymer:curing agent solution. Mix thoroughly and remove air bubbles by placing the mixture in a vacuum dessicator. This volume of PDMS is sufficient for a 2- or 3-in. silicon master in a standard 100-mm Petri dish.

2. Pour degassed PDMS mixture onto silicon master.
3. Bake at 60°C for 2 h.
4. Carefully peel hardened PDMS stamp from silicon master. Silicon masters can be reused indefinitely.
5. Treat stamp for 5 min in UV/ozone cleaner. Stamps should be used to make agarose wells on the same day as UV/ozone treatment (*see* **Note 1**).

3.3. Construction of Agarose Wells

The goal of this procedure is to fill the channels between the PDMS stamp and the glass slide with agarose. When the stamp is removed, the agarose "walls" surround glass-bottomed "wells," into which the cells will later be seeded. There are two major obstacles to filling PDMS channels with agarose: (1) the PDMS channels are quite hydrophobic and prevent the flow of aqueous solutions through them, and (2) aqueous agarose solutions are quite viscous, have a high surface tension, and do not flow very readily. The first obstacle is addressed (**Subheading 3.2., step 5**) by treating the PDMS stamp in the UV/ozone cleaner, which modifies the surface of the stamp to render it more hydrophilic and helps the solution flow through the channels. The second obstacle is addressed by adding ethanol to the agarose solution to decrease both its viscosity and surface tension. The procedure described below was arrived at empirically in our laboratory, and certain details might need to be altered depending on experimental conditions.

1. Prepare a solution of 1% agarose in distilled water. Heat to 80°C. The stock solution of 1% agarose can be reused, but should be stored in a tightly sealed container to prevent evaporation or contamination.
2. Mix six parts 1% agarose solution with four parts 200-proof ethanol. Because ethanol evaporates readily, the agarose/ethanol solution should be prepared fresh for each batch of microwells (*see* **Note 2**).
3. Seal UV/ozone-treated stamp against Superfrost slide, feature-side down.
4. Pipet a drop of the hot agarose/ethanol solution against the side of the sealed stamp. Allow the solution to perfuse through the channels formed between the stamp and the slide. The volume of agarose/ethanol solution required per sample will depend on the size of the stamp. Adequate perfusion can be obtained with stamps approx 0.5 cm wide.
5. Dry the slides by placing in a vacuum dessicator for approx 10 min, or until dry. During this time the agarose will gel and the solvent will evaporate, leaving a thin film of agarose in the desired pattern on the glass slide (*see* **Note 3**).
6. Carefully remove the stamp from the slide with fine-tipped forceps. During removal, avoid any lateral movement of the stamp, which may smear the agarose film and deform or destroy the microwells (in other words, pull the stamp straight up). Each microwell will consist of a base of bare glass surrounded by walls of agarose and can be easily inspected for defects using a standard microscope (**Fig. 2A**).

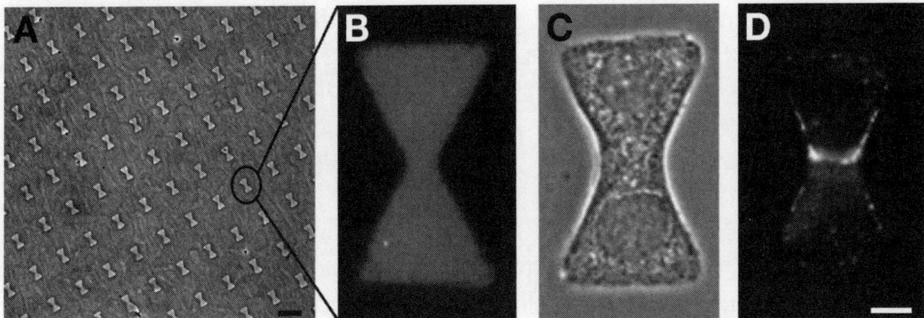

Fig. 2. Bowtie-shaped agarose microwells restrict protein adsorption, cell attachment and spreading, and cell–cell adhesion. Shown are **(A)** phase contrast image of agarose wells, **(B)** fluorescence image of staining for fibronectin, **(C)** phase contrast image of cells, and **(D)** β-catenin in bowtie-shaped microwells. **A:** Bar = 50 µm; **B–D:** Bar = 10 µm.

3.4. Coating and Cell Culture

After construction of the agarose microwells, cells can be seeded directly into the untreated glass wells or the wells may first be coated with a specific ECM protein to which the cells can subsequently adhere. For endothelial cells, we typically coat the wells with a solution of fibronectin. Because agarose is a hydrogel, it is relatively resistant to protein adsorption, so fibronectin adsorbs only to the bare glass bases of the microwells *(5)* (**Fig. 2B**). Because agarose resists adsorption of ECM proteins necessary for cell adhesion, it consequently resists cell attachment; when cells are seeded onto the substratum, they attach exclusively to the glass-bottomed wells (*see* **Note 4**).

1. Place agarose-coated slides in 70% ethanol solution to sterilize for cell culture.
2. Rinse twice with PBS.
3. Coat with 25 µg/mL solution of fibronectin in PBS for 1–2 h at room temperature.
4. Rinse twice with PBS.
5. Plate cells on agarose-coated slides in normal culture media. Allow cells to adhere in wells. Rinse to remove excess cells (**Fig. 2C,D**).

3.5. Experimental Analysis

We have found that a typical substratum will yield several hundred bowtie-shaped cultures, some of which contain a pair of contacting cells, and some of which contain a single cell without a neighbor. Each substratum sample therefore consists of a subpopulation of cells (the single cells) that acts as an internal control. However, because of the small number of cells and the heterogeneity of the bowties on a single substratum, experimental analysis is limited primarily to *in situ* techniques. We have used the bowtie microwells to examine the effects of cell–cell contact on proliferation by staining for BrdU incorporation *(5,19)* and on cell–ECM adhesion by immunofluorescence analysis *(20)*. One should also be able to combine the bowtie microwells with *in situ* hybridization or *in situ* reverse transcription–polymerase chain reaction (RT/PCR) to study changes in gene expression. Theoretically, one should be able to scale up the technique; larger patterned areas or a higher percentage of patterned pairs of cells would allow one to obtain enough material for bulk population analysis (via Western blotting, etc.).

The method is adaptable for different types of cells. In addition to a number of types of endothelial cells, we have successfully cultured smooth muscle cells, fibroblasts, and epithelial cells in the bowtie microwells (*see* **Note 5**). The agarose bowties are best suited for experiments of short duration (less than a few days), because the agarose layer tends to detach from the glass over extended periods in aqueous solutions, such as culture media *(21)* (*see* **Note 6**).

4. Notes

1. Stamps can be reused four to five times.
2. If the final ethanol concentration is too high, agarose will precipitate out of solution and the film is no longer nonadhesive.
3. The agarose layer will detach from the slide if it is not first dried completely.
4. Because of chemical modifications, low-melt agarose is not resistant to protein adsorption. Thus, cells would adhere everywhere to such a substratum.
5. Some of the patterning parameters may need to be altered when working with different types of cells. We have found that longer drying times are needed to increase the stiffness of the agarose gel for some cell lines (such as normal rat kidney epithelial cells).
6. For experiments that require patterning cells over longer time scales, we have developed an alternative method of constructing wells with walls of nonadhesive polyacrylamide gel *(21)*. These substrata will maintain their pattern for at least 2 mo.

Acknowledgments

This work was supported by grants from the NIH, the Whitaker Foundation, and the Defense Advanced Research Projects Agency. C.M.N. acknowledges financial support from the Whitaker Foundation, and W.F.L. was supported by the National Science Foundation.

References

1. Folch, A. and Toner, M. (2000) Microengineering of cellular interactions. *Annu. Rev. Biomed. Eng.* **2,** 227–256.
2. Chen, C. S., Mrksich, M., Huang, S., Whitesides, G. M., and Ingber, D. E. (1997) Geometric control of cell life and death. *Science* **276,** 1425–1428.
3. McBeath, R., Pirone, D. M., Nelson, C. M., Bhadriraju, K., and Chen, C. S. (2004) Cell shape, cytoskeletal tension, and RhoA regulate stem cell lineage commitment. *Dev. Cell* **6,** 483–495.

4. Bhatia, S. N., Balis, U. J., Yarmush, M. L., and Toner, M. (1999) Effect of cell–cell interactions in preservation of cellular phenotype: cocultivation of hepatocytes and nonparenchymal cells. *FASEB J.* **13,** 1883–1900.
5. Nelson, C. M. and Chen, C. S. (2002) Cell–cell signaling by direct contact increases cell proliferation via a PI3K-dependent signal. *FEBS Lett.* **514,** 238–242.
6. Nelson, C. M. and Chen, C. S. (2003) VE-cadherin simultaneously stimulates and inhibits cell proliferation by altering cytoskeletal structure and tension. *J. Cell Sci.* **116,** 3571–3581.
7. Huang, S. and Ingber, D. E. (1999) The structural and mechanical complexity of cell-growth control. *Nat. Cell Biol.* **1,** E131–138.
8. Folkman, J. and Moscona, A. (1978) Role of cell shape in growth control. *Nature* **273,** 345–349.
9. Singhvi, R., et al. (1994) Engineering cell shape and function. *Science* **264,** 696–698.
10. Holley, R. W. and Kiernan, J. A. (1968) "Contact inhibition" of cell division in 3T3 cells. *Proc. Natl. Acad. Sci. USA* **60,** 300–304.
11. Stoker, M. G. and Rubin, H. (1967) Density dependent inhibition of cell growth in culture. *Nature* **215,** 171–172.
12. Steinberg, M. S. and Takeichi, M. (1994) Experimental specification of cell sorting, tissue spreading, and specific spatial patterning by quantitative differences in cadherin expression. *Proc. Natl. Acad. Sci. USA* **91,** 206–209.
13. Gonzalez-Mariscal, L., Chavez de Ramirez, B., and Cereijido, M. (1985) Tight junction formation in cultured epithelial cells (MDCK). *J. Membr. Biol.* **86,** 113–125.
14. Nakamura, T., Yoshimoto, K., Nakayama, Y., Tomita, Y., and Ichihara, A. (1983) Reciprocal modulation of growth and differentiated functions of mature rat hepatocytes in primary culture by cell—cell contact and cell membranes. *Proc. Natl. Acad. Sci. USA* **80,** 7229–7233.
15. Takeichi, M. (1977) Functional correlation between cell adhesive properties and some cell surface proteins. *J. Cell Biol.* **75,** 464–474.
16. Kovacs, E. M., Ali, R. G., McCormack, A. J., and Yap, A. S. (2002) E-cadherin homophilic ligation directly signals through Rac and phosphatidylinositol 3-kinase to regulate adhesive contacts. *J. Biol. Chem.* **277,** 6708–6718.
17. Levenberg, S., Katz, B. Z., Yamada, K. M., and Geiger, B. (1998) Long-range and selective autoregulation of cell–cell or cell-matrix adhesions by cadherin or integrin ligands. *J. Cell Sci.* **111 (Pt. 3),** 347–357.
18. Lambert, M., Padilla, F., and Mege, R. M. (2000) Immobilized dimers of N-cadherin-Fc chimera mimic cadherin-mediated cell contact formation: contribution of both outside-in and inside-out signals. *J. Cell Sci.* **113,** 2207–2219.
19. Liu, W. F., Nelson, C. M., Pirone, D. M., and Chen, C. S. (2006) E-cadherin engagement stimulates proliferation via Rac1. *J. Cell Biol.* **173,** 431–441.
20. Nelson, C. M., Pirone, D. M., Tan, J. L., and Chen, C. S. (2004) Vascular endothelial-cadherin regulates cytoskeletal tension, cell spreading, and focal adhesions by stimulating RhoA. *Mol. Biol. Cell* **15,** 2943–2953.
21. Nelson, C. M., Raghavan, S., Tan, J. L., and Chen, C. S. (2003) Degradation of micropatterned surfaces by cell-dependent and -independent processes. *Langmuir* **19,** 1493–1499.

2

Separation of Cell–Cell Adhesion Complexes by Differential Centrifugation

Roger Vogelmann and W. James Nelson

Summary

The number of proteins found associated with cell–cell adhesion substructures is growing rapidly. Based on potential protein–protein interactions, complex protein networks at cell–cell contacts can be modeled. Traditional studies to examine protein–protein interactions include co-immunoprecipitation or pull-down experiments of tagged proteins. These studies provide valuable information that proteins can associate directly or indirectly through other proteins in a complex. However, they do not clarify if a given protein is part of other protein complexes or inform about the specificity of those interactions in the context of adhesion substructures. Thus, it is not clear if models compiled from these types of studies reflect the combination of protein interactions in the adhesion complex in vivo for a specific cell type. Therefore, we present here a method to separate cell–cell contact membrane substructures with their associated protein complexes based on their buoyant behavior in iodixanol density gradients. Analysis of 16 proteins of the apical junctional complex (AJC) in epithelial Madin–Darby canine kidney cells revealed a more simple organization of the AJC adhesion complex than that predicted from the combination of all possible protein–protein interactions defined from co-immunoprecipitation and pull-down experiments.

Key Words: Cell polarity; tight junction; adherens junction; protein complex; Ig superfamily receptors; apical junctional complex; density gradient.

1. Introduction

Cell–cell adhesion complexes are important regulators of cellular functions. The apical junctional complex (AJC) in epithelia regulates cell–cell adhesion between neighboring cells, structural and functional integrity of the epithelial barrier, contractile forces during morphogenesis and wound healing, cell proliferation, cell differentiation, and cell polarity *(1,2)*. In order to understand

From: *Methods in Molecular Biology, vol. 370:*
Adhesion Protein Protocols, Second Edition
Edited by: A. S. Coutts © Humana Press Inc., Totowa, NJ

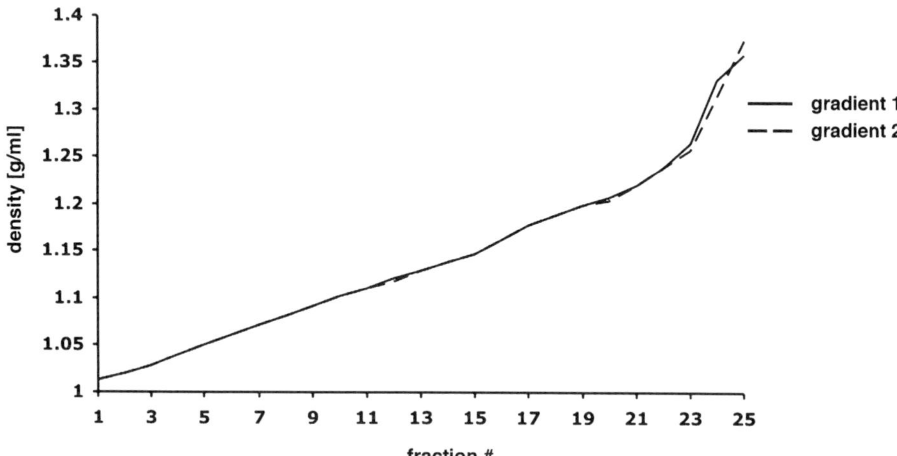

Fig. 1. Density of 10–20–30% iodixanol density gradients. Density δ was determined for each fraction by measuring the refractive index (η) of each fraction [$\delta = (\eta *$ 3.443) - 3.599]. Densities for two different density gradients are shown (gradient 1 and gradient 2).

how adhesion complexes carry out their functions, it is important to know the identity of proteins that localize to the AJC and the nature of the protein–protein interactions as a basis for understanding these different functions. In vitro studies have revealed a large number of proteins and their potential binding partners at the AJC, which has led to models of complex protein networks *(3–5)*. However, these studies do not clarify if all of these protein–protein interactions occur at the same time in a given cell type, or at a particular stage of assembly of the AJC, or if proteins are separated into different AJC substructures. Whether or not two proteins are in the same complex is critical for the understanding of how adhesion complexes mechanistically regulate multiple functions of the AJC.

Membrane-associated protein complexes can be separated using iodixanol density gradients *(6,7)*. Iodixanol is an aqueous solution that is iso-osmotic up to a density of 1.32 g/mL and forms self-generating gradients in 1–3 h. It has a very low toxicity toward biological material, and enzyme assays can be carried out in its presence.

Yeaman et al. developed a 10–20–30% iodixanol density gradient to study protein trafficking of the Sec 6/8 protein complex in epithelial cells *(8)*. This type of gradient exhibits a linear increase in density over almost the entire length of the gradient, and membrane particles are separated in the gradient based on their buoyant characteristics (**Fig. 1**). Cells are broken mechanically in a buffer lacking detergent to preserve membranes in the lysate. After loading the cell lysate at the bottom of the gradient, membranes float up as the iodixanol gradient self-generates; the separation of different membranes depends on their lipid content and the concentration of associated proteins. After centrifugation, fractions can be sampled from the top of the gradient and analyzed using standard sodium dodecyl sulfate–polyacrylamide gel electrophoresis (SDS-PAGE) and immuoblot staining for proteins of interest. Proteins separating in different fractions of the gradient are in neither the same protein complex nor the same AJC substructure. However, proteins that separate in the same fractions may interact with each other in the same complex, which can be further tested by immuno-precipitiation of proteins from the separated membranes.

We utilized the 10–20–30% iodixanol density gradient method to analyze 16 proteins of the AJC in epithelial Madin–Darby canine kidney (MDCK) cells *(9)*. Our study revealed that the organization of the AJC is simpler than the predicted model based on protein–protein interaction studies. Many of the analyzed proteins are separated into distinct subcomplexes or may interact only transiently during AJC assembly, whereas others may be irrelevant to the formation and maintenance of the AJC in MDCK cells. Iodixanol density gradients are, therefore, a useful biochemical strategy to isolate cell–cell adhesion complexes as a starting point to further study the composition of protein complexes in these structures.

2. Materials

2.1. Cell Culture and Calcium-Switch Experiment

1. Dulbecco's modified Eagle's medium (DMEM) (Invitrogen-Gibco, Carlsbad, CA) containing 10% fetal bovine serum (FBS; Atlas Biologicals, Fort Collins, CO) and 5 mL of 100X penicillin, streptomycin, and kanamycin (PSK) antibiotics mix. To prepare PSK antibiotics mix, dissolve 6.1 g kanamycin sulfate (100 mg/mL; Invitrogen-Gibco), 1.5 g penicillin "G" sodium (50 u/mL, 1650 u/mg; Invitrogen-Gibco), and 2.5 g streptomycin sulfate (50 mg/mL; Sigma-Aldrich, St. Louis, MO) in 500 mL of phosphate-buffered saline (PBS). Filter through a 0.22-μm filter unit to sterilize, dispense 50 mL per 100-mL bottle, and store at −20°C for up to 6 mo.
2. Washing buffer: dissolve 48 g NaCl, 2.4 g KCl, 6 g glucose (dextrose, monohydrate), 2.1 g NaHCO$_3$ in 5.95 L dH$_2$O and bring to a final volume of 6 L. Add 1.2 g ethylene diamine tetraacetic acid (will take a while to dissolve), and filter-sterilize using a 0.22-μm filter unit. Aliquot 500 mL each into 500-mL glass tissue culture bottles, and store at 4°C.
3. Washing buffer + trypsin: for trypsin stock solution, dissolve 6.25 g trypsin (Difco, BD, Franklin Lakes, NJ) in 250 mL of washing buffer. Stir 20 min at room temperature (solution will remain cloudy). Centrifuge for 30 min at 10,000 rpm (12,064g) 4°C in JA-20 rotor (Beckman Instruments Inc., Palo Alto, CA). Filter-sterilize supernatant using 0.22-μm filter unit, and discard pellet. Aliquot 12.5 mL each of sterile trypsin solution into 20 sterile 50-mL blue cap tubes, store at −20°C. For trypsin working solution (0.0625%), add one tube of sterile trypsin stock solution to one bottle (500 mL) of sterile washing buffer, and mix well.

4. Low-calcium medium (for 6 L): 2.4 g KCl, 1.1989 g MgSO$_4$·7H$_2$O, 35.777 g NaCl, 6.0 g D-glucose (dextrose, monohydratous), 0.06 g phenol red, 0.756 g L-arginine HCl, 0.18774 g L-cystine 2HCl, 1.7752 g L-glutamine, 0.252 g L-histidine HCl·2H$_2$O, 0.312 g L-isoleucine, 0.312 g L-leucine, 0.435 g L-lysine HCl, 0.192 g L-phenylalanine, 0.288 g L-threonine, 0.06 g L-tryptophan, 0.31188 g L-tyrosine, 0.275 g L-valine (*see* **Note 1**), 60 mL 100X minimum essential medium (MEM) vitamin solution (Invitrogen-Gibco), 6 g NaHCO$_3$, 15.618 g hydroxyethylpiperazine ethanesulfonic acid (HEPES) NaCl, 0.168 mL 100X CaCl$_2$ (dissolve 2.65 g CaCl$_2$·2H$_2$O in 90 mL of dH$_2$O and bring up to final volume of 100 mL; filter-sterilize with 0.22-μm filter unit, aliquot 20 mL each into 50-mL blue cap tubes, store at –20°C), 0.84 g NaH$_2$PO$_4$·H$_2$O. Adjust pH to 7.0 using concentrated HCl, adjust volume with dH$_2$O to 6 L, filter-sterilize, and store media (500 mL/bottle) at 4°C. Add 10% dialyzed FBS, 5 mL of 100X methionine, and 5 mL of 100X PSK for the working solution.
5. Dialyzed FBS: entire procedure is done at 4°C. Prepare 4 L of Tris-saline (10 mM Tris-HCL pH 7.5, 120 mM NaCl; dialysis solution) and cool to 4°C. Divide 500 mL of FBS into five or six sections of 3/4-in.-wide dialysis tubing that has been rinsed with dH$_2$O. Put FBS-containing dialysis tubes into one 4-L beaker of Tris-saline and stir gently. Change Tris-saline solution after 24 h and repeat three times. Filter-sterilize FBS using a 500-mL 0.22-μm filter unit, aliquot into 50-mL sterile blue cap tubes, and store at –20°C.
6. 100X Methionine: dissolve 0.15 g of methionine in 90 mL of dH$_2$O and bring to a final volume of 100 mL. Filter-sterilize in 0.22-μm filter unit, and store at –20°C.
7. Collagen solution: commercial collagen type I solutions can be used. For instructions on how to prepare your own collagen solution, please visit our web page at http://nelsonlab.stanford.edu/lab/labbible.html#_Toc50366772.
8. 75-mm-Well, 0.4-μm polycarbonate membranes (Transwell® filters; Costar Corp., Cambridge, MA).

2.2. Cell Surface Biotinylation

1. Sulfo-NHS-Biotin (Pierce Biotechnology Inc., Rockford, IL) at a final concentration of 0.5 mg/mL in ice-cold Ringers solution. Prepare fresh before each experiment and use immediately.
2. Ringers solution: 10 mM HEPES NaCl, pH 7.4, 154 mM NaCl, 7.2 mM KCl, 1.8 mM CaCl$_2$.

2.3. Cell Fractionation in Iodixanol Gradients

1. Dithiobis(succinimidylpropionate) (DSP; Pierce Biotechnology Inc.) stock: 20 mg/mL in dimethylsulfoxide (DMSO) (always make fresh). Dilute DSP stock 1:100 in Ringers solution at room temperature immediately prior to use.
2. Quenching buffer: 120 mM NaCl, 10 mM Tris-HCl, pH 7.4, 50 mM NaH$_4$Cl.
3. Homogenization buffer 10X stock: 200 mM HEPES·KOH, pH 7.2 (*see* **Note 2**), 900 mM K-acetate, 20 mM Mg-acetate. Filter-sterilize in 0.22-μm filter unit, and store at 4°C.

4. Homogenization buffer I (make fresh for each experiment): prepare 1 mL for each 75-mm filter. For one gradient (8- × 75-mm filters): 7.97 mL dH$_2$O, 900 µL homogenization buffer 10X, 90 µL sucrose (2.5 M stock), 30 µL Pefabloc (0.1 M stock; Hoffmann-La Roche Inc., Nutley, NJ), 8 µL L/A/P (5 mg/mL leupeptin, 5 mg/mL antipain, 5 mg/mL pepstatin in DMSO; Hoffmann-La Roche) (*see* **Note 3**).
5. 15-mL Conical polypropylene tubes (*see* **Note 4**).
6. Homogenization buffer II (make fresh for each experiment): prepare 3 mL for one gradient. 2.63 mL dH$_2$O, 300 µL homogenization buffer 10X, 30 µL sucrose (2.5 M stock), 30 µL Pefabloc (0.1 M stock), 8 µL aprotinin (Sigma-Aldrich), 8 µL L/A/P (*see* **Notes 3** and **5**).
7. Homogenization buffer III (make fresh for each experiment): prepare 13 mL for two gradients. 11.57 mL dH$_2$O, 1.3 mL homogenization buffer 10X, 130 µL sucrose (2.5 M stock).
8. Ball-bearing homogenizer, fitted with 0.3747-in. stainless steel ball (**Fig. 2**) (*see* **Note 6**).
9. 5-mL syringes with Luer-Lok™ tip (BD, Franklin Lakes, NJ), 23G1 needles.
10. Iodixanol: Opti-prep® (Nycomed, Oslo, Norway).
11. Quick-Seal® ultracentrifuge tube, 5/8 × 3 in. (16 × 76 mm) (Beckman Instruments Inc.).

2.4. SDS-PAGE and Immunoblotting

1. 4X SDS sample buffer: for 20 mL solution: 1.6 g SDS, 3.2 mL 1 M Tris-HCl pH 6.8, 6 mL glycerol (100%), bromophenol blue, add dH$_2$O up to 20 mL. Store at room temperature. Before use, add 1 mL of 1 M dithiothreitol (store at −20°C) to 4 mL 4X SDS-sample buffer.
2. 10% SDS stock solution, 1 M Tris-HCl, pH 8.7 (for separating gel), 1 M Tris-HCl, pH 6.8 (for stacking gel). Store at room temperature.
3. 30% Acrylamide/0.8% bisacrylamide solution (National Diagnostics, Atlanta, GA) and N,N,N,N'-tetramethyl-ethylenediamine (TEMED; Bio-Rad, Hercules, CA). Acrylamide is a neurotoxin when unpolymerized; see manufacturer's safety instructions before use. 10% Ammonium persulfate (APS; Bio-Rad) in dH$_2$O. Store all solutions at 4°C.
4. SDS running buffer (10X): 250 mM Tris, 1.92 M glycine, 1% (w/v) SDS. Store at room temperature. Dilute 1:10 in dH$_2$O before use.
5. Prestained molecular-weight markers.
6. Western blot transfer buffer: 25 mM Tris (do not adjust pH), 192 mM glycine, 20% (v/v) methanol.
7. Protran® 0.45-µm nitrocellulose membranes (Schleicher & Schuell, Keene, NH), Immobilon-FL polyvinylidene fluoride (PVDF) Odyssey membrane (Millipore, Billerica, MA) (*see* **Note 7**).
8. Washing buffer: Tris-buffered saline with Tween: 120 mM NaCl, 10 mM Tris-HCl, pH 7.4, + 0.1% Tween-20.
9. Blocking buffer: LI-COR blocking buffer (LI-COR, Lincoln, NE) in PBS (1:1).

10. Primary antibodies and fluorescently labeled secondary antibodies (1:30,000) are diluted in washing buffer.
11. Secondary antibodies: anti-mouse Alexa Fluor® 680, Alexa® Fluor 680 coupled streptavidin (Molecular Probes, Eugene, OR), and anti-rabbit IRDye® 800 (Rockland Immunochemicals, Gilbertsville, PA).
12. Odyssey® Infrared Imaging System (LI-COR).

3. Methods

3.1. Cell Culture and Calcium-Switch Experiment

The number of cells required for iodixanol gradient separation of the AJC will vary depending on the detection limit of proteins in membrane complexes following centrifugation. Cell–cell adhesion is synchronized by a calcium-switch experiment, which allows adhesion complexes of the AJC to be isolated at different stages of assembly.

1. MDCK II cells are cultured in DMEM + 10% FBS + 1% PSK at 37°C/5% CO_2 in air.
2. For cell passaging, a 150-mm TC culture dish is washed with 8 mL washing buffer and incubated with 8 mL washing buffer + trypsin for 15–30 min at 37°C/5% CO_2 in air. Add 3 mL of complete DMEM media after cells have lifted off the plate to inactivate trypsin; centrifuge at 750g for 5 min and resuspend pellet in appropriate amount of media. All work is performed under sterile conditions.
3. Calcium-switch experiment: pass cells at low density (2×10^6 cells/150 mm TC culture dish) on 2 consecutive days. For one gradient use 10 150-mm TC culture dishes, trypsinize cells as described above on day 1 and day 2, and divide all cells ($\sim 2 \times 10^8$ cells) among 10 new 150-mm TC culture dishes.
4. Collagen coat Transwell filters: add 4 mL of collagen working solution to a 75-mm filter and incubate for 1–2 min at room temperature. Remove collagen solution and dry filters under UV light for 30–60 min at room temperature to sterilize filters (*see* **Note 8**).
5. After trypsinization at day 3, plate 2×10^7 cells on collagen-coated 75 mm 0.4-µm polycarbonate membranes in low-calcium media (5 µM Ca^{2+}) for 30 min to allow attachment. Plate eight 75-mm Transwell filters per density gradient.
6. Transfer cells into regular DMEM (1.8 mM Ca^{2+}) for 3 h, 6 h, 9 h, 12 h, 3 d, and 10 d. For time-points 3 and 10 d, replace media every 24 h with fresh DMEM.

3.2. Cell Surface Biotinylation

For some experiments it can be helpful to identify transmembrane proteins of the plasma membrane (*7,9*). Cell surface biotinylation allows the identification of transmembrane proteins in different plasma membrane domains. Because of the barrier function of tight junctions, apical and basal–lateral membranes can be biotinylated separately using the Transwell filter system. This step can be omitted if identification of transmembrane proteins is not necessary.

1. Grow MDCK II cells as a confluent monolayer for 3 d on 75-mm Transwell filters.
2. Perform **steps 2–5** in the cold room or on ice. Wash three times in ice-cold Ringers solution.
3. Add Sulfo-NHS-Biotin (0.5 mg/mL) in ice-cold Ringers solution to apical or basal or both compartments of the filter to study proteins in the apical or basal or both membrane domains, respectively (4 mL in apical and 7 mL in basal compartment).
4. Incubate cells twice for 20 min at 4°C with gentle rocking. In between, remove solutions after first 20 min and add fresh Ringers solution with or without Biotin.
5. Wash three times with Ringers solution and proceed with cell fractionation protocol.

3.3. Cell Fractionation in Iodixanol Gradients

3.3.1. Cell Fractionation

During breaking of cells and fractionating of membranes, it is possible that protein complexes can dissociate. In order to preserve existing adhesion structures, we suggest chemically crosslinking proteins prior to mechanical breakage of cells.

1. Wash cells three times with ice-cold Ringers solution in a cold room or on ice.
2. Add 6 mL of freshly prepared 200 µg/mL DSP in Ringers solution to the apical compartment and incubate for 20 min at room temperature with gentle rocking.
3. Perform the following steps in a cold room or on ice: quench the crosslinking reaction with five washes in quenching buffer.
4. Wash cells three times in ice-cold Ringers solution and add 1 mL of homogenization buffer I to the apical compartment. Scrape cells from filter surface and pool cells from eight Transwell filters in 15 mL conical polypropylene tube.
5. Pellet cells by centrifugation at 228*g* for 10 min at 4°C.
6. Resuspend pellet in 2 mL of homogenization buffer II using a 5-mL syringe with a screw cap, 23G1 needle.
7. Homogenize cells in ball-bearing homogenizer fitted with a stainless steel ball bearing (**Fig. 2**); pass cell homogenate back and forth 20 times; flush homogenizer with remaining 1 mL of homogenization buffer II (*see* **Note 9**).
8. Spin homogenate at 930*g* for 10 min at 4°C to pellet nuclei and unbroken cells.

3.3.2. Iodixanol Gradient

After mechanical breakage of cells, membranes are separated by centrifugation in a three-phase 10–20–30% iodixanol density gradient.

1. Prepare 10, 20, and 30% iodixanol solutions according to **Table 1** (*see* **Note 10**).
2. Measure density of each solution with a refractometer. Adjust solutions accordingly to a refractive index (η) of 1.359 for 10%, 1.373 for 20%, and 1.387 for 30% solution (all at room temperature).
3. Add all of the 30% solution containing the cell homogenate to the bottom of the centrifuge tube with a Pasteur glass pipet. Overlay with 4.5 mL of the 20% solution, and fill the tube completely with an overlay of 10% solution (*see* **Note 11**).

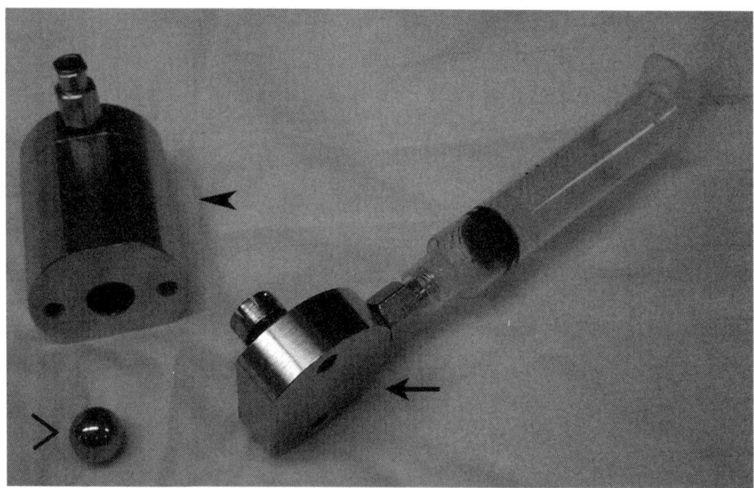

Fig. 2. Ball-bearing homogenizer, designed by Varian Physics Department, Stanford University, CA. The front part (←) is detached to demonstrate a ball bearing (>) and steel cylinder (◄). For proper function, the ball bearing is placed inside the cylinder and the front and back assembled with a screw. Syringes are attached to both syringe adaptors as shown for the front part.

Table 1
Solutions for 10–20–30% Iodixanol Gradient

Solution (%)	Iodixanol, 60% w/v (mL)	Homogenization buffer III (mL)	Postnuclear supernatant (mL)
10	1.25	3.25	—
20	2	2.7	—
30	2.95	—	2.25

4. After the tubes are balanced within 0.1 g of each other, seal the tube neck using standard Beckman equipment for tube sealing.
5. Centrifuge gradients in a VTI 65.1 rotor at 61,000 rpm (350,000g) for 3 h 10 min (decel = 5) in L8-80M Beckman Ultracentrifuge (Beckman Instruments Inc.).
6. After centrifugation, open tubes and collect 0.5-mL fractions from the top of the gradient (*see* **Note 12**).
7. Measure the refractive index (η) of each fraction using a refractometer and determine density δ in g/mL [$\delta = (\eta * 3.443) - 3.599$]. Plot the density δ against the fraction number; the density should increase linearly almost over the entire gradient (**Fig. 1**).

Some membranes associated with larger protein complexes remain at the bottom of the gradient and do not float up in the 10–20–30% floating iodixanol gradient. However, these membranes can be separated in a second iodixanol gradient *(9)*:

1. Separate membranes in a 10–20–30% floating iodixanol gradient as described above.
2. Combine fractions 19–22 (or other fractions of interest) and dilute to a 10% iodixanol solution.
3. Form a 10–20–30% iodixanol gradient as described above using homogenization buffer III for the 30% solution and the 10% iodixanol solution containing fractions 19–22 from the previous gradient.
4. Proceed as described in **steps 3–7** above.

3.4. SDS-PAGE and Immunoblotting

For the analysis of proteins in the iodixanol gradient, any standard SDS-PAGE and immunoblotting technique can be used. We describe here the method that works well for this laboratory.

1. Add 170 µL of SDS sample buffer to 500 µL of fraction sample, boil for 10 min, and store at −80°C (*see* **Note 13**).
2. Most proteins can be analyzed in linear 14, 10, or 7.5% SDS-polyacrylamide gels. Choose the appropriate separating gel for the molecular mass of your protein of interest. Mix 2.98 mL 1 *M* Tris-HCl pH 8.7, 80 µL 10% SDS, 4 µL TEMED, and 27 µL APS. Add 1.06 mL dH$_2$O and 3.7 mL acrylamide/*bis* for 14%, 2.2 mL dH$_2$O and 2.7 mL acrylamide/*bis* for 10%, or 2.8 mL dH$_2$O and 2 mL acrylamide/*bis* for 7.5% separating gel, respectively. After polymerization of the separating gel, add the stacking gel (3.8 mL dH$_2$O, 0.64 mL 1 *M* Tris-HCl pH 6.8, 50 µL 10% SDS, 0.5 mL acrylamide/*bis*, 5 µL TEMED, and 25 µL 10% APS). This gel recipe is for two minigels; adjust accordingly for different styles of slab gels.
3. After loading gels with molecular-weight marker and samples, separate proteins in SDS running buffer. Transfer proteins onto nitrocellulose or PVDF membranes using a standard Western blot transfer protocol (*see* **Note 14**).

The quantification of the protein signal after immunoblotting is important to compare sedimentation profiles of different proteins (**Fig. 3**). We use an Odyssey Infrared Imagining System to quantify protein signals, but other quantification methods can be used equally well.

1. Block proteins with LI-COR blocking buffer in PBS (1:1) for 1 h at room temperature.
2. Immunostain proteins with appropriate primary antibodies diluted in washing buffer for 1 h at room temperature.
3. Rinse membranes twice followed by 15-min and three 5-min washes in washing buffer.
4. Incubate membranes with appropriate fluorescently labeled secondary antibody diluted at 1:30,000 in washing buffer for 30 min at room temperature. *It is important to light-protect your membranes in this and the following steps!*

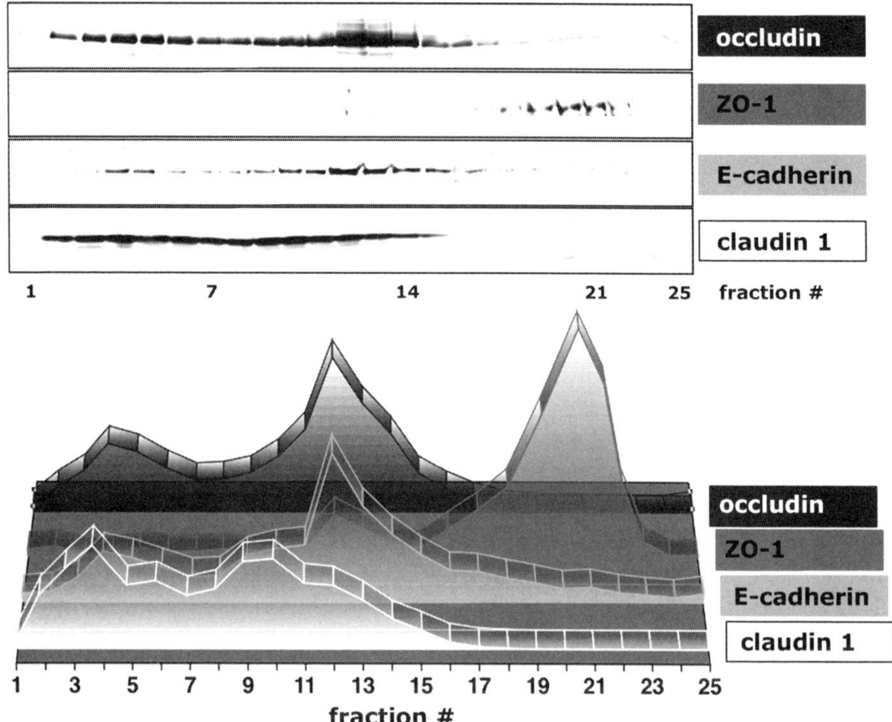

Fig. 3. Proteins of the apical junctional complex (AJC) in polarized epithelia (3 d after cell–cell adhesion) separated in iodixanol gradients. Madin–Darby canine kidney (MDCK) cells were grown as confluent monolayers on Transwell® filters for 3 d after cell–cell adhesion and membranes separated in 10–20–30% iodixanol gradients. The top panel shows immunoblot analysis of indicated proteins. Signal intensity for each protein band was determined as integrated intensity (counts/mm^2) and expressed as a percentage of the sum of integrated intensities in fractions 1–25 (*see* text for details). In the three-dimensional graph, the *y*-axis (arbitrary units) is omitted to increase the clarity of the graphic display.

5. Rinse membranes twice followed by three 5-min washes in washing buffer in the dark. Rinse membranes twice with PBS, and store in PBS at 4°C (protect from light).
6. Scan membranes with Odyssey Infrared Imaging System at 680 and 800 nm.
7. Determine the amount of protein per fraction with Odyssey software 1.2. Express data as integrated intensity of a specified area and correct for background using the appropriate background subtraction method (for details, please refer to user guide, version 1.2 for Odyssey Infrared Imaging System, Chapters 8 and 12). For each protein, the integrated intensities of fractions 1–25 in one gradient can be summed up and the amount in one fraction can be expressed as percent of the sum of fractions 1–25 (*see* **Fig. 3**).

4. Notes

1. Amino acids need to mix at least 1 h to dissolve.
2. Use HEPES·KOH instead of the more common HEPES·NaCl because the salt content is critical for density gradient characteristics. Make 1 M HEPES·KOH stock and adjust pH to 7.2 using KCl. Sterile filter (0.22-μm filter unit) and store at 4°C.
3. Dilute 5 mg leupeptin, 5 mg antipain, and 5 mg pepstatin in 1 mL of DMSO, aliquot 50 μL per vial and store at −20°C. Note that any commercial available proteinase inhibitor cocktail can be used according to the manufacturer's instructions.
4. Cell fragments stick less to polypropylene tubes.
5. Use precooled solutions (including dH_2O) or make sure that homogenization buffers are at 4°C before use.
6. The method for breaking cells is critical. We use a ball-bearing homogenizer designed by Varian Physics Department, Stanford University.
7. Immobilon-FL PVDF membranes are preferred for less abundant proteins (higher recovery) and if stripping of blots is desired. In contrast to Immobilon-FL, other PVDF membranes from Millipore have a high background in Odyssey Infrared Imaging System.
8. Collagen coating is critical for rapid cell attachment. Without coating, cells require more than 3 h to attach to polycarbonate membranes.
9. Have about 2 mL of air in the second empty syringe, which is already attached to the ball-bearing homogenizer. Add about 2 mL of air to the syringe containing the cell suspension before attaching to the homogenizer. Move cells completely from one syringe to the other. Frequent small back-and-forth movements while directing the net flow in one direction are very helpful. Clean the ball-bearing homogenizer between different cell lysates by flushing the homogenizer with 4 mL dH_2O; remove excessive fluid in the homogenizer by passing air through it.
10. Prepare solutions for an even number of gradients (balance for centrifuge). Table 1 gives information for one gradient. Prepare 10 and 20% solutions as one pool, respectively, for the number of gradients you are running (multiply numbers in Table 1). After adjustment for refractive index, divide 20% solution in 4.5-mL aliquots.
11. Balance tubes after each step. Tubes should be balanced within 0.1 g of each other after all solutions have been added. Add 20 or 10% solutions very slowly to avoid mixing with 30 or 20% solution, respectively. Sometimes it can be difficult to get the last air bubble out of the tube to fill it completely including the neck. It can be helpful to use a 1-mL syringe and a 26G1/2 needle instead of the Pasteur pipet for the last drops.
12. To open up the ultracentrifuge tubes, use a fresh razor blade and a hemostat. Fractions can be collected by different techniques. Use a small peristaltic pump and an automatic collector while keeping the tip of the collecting tube at the surface of the gradient slowly moving downward. Alternatively, collect initially 100-μL fractions with a P100 or P200 pipet aid and combine them to 500 μL. After the third 500-μL fraction, open the tube completely using a hot razor blade and collect 500-μL fractions with a P1000 pipet aid, moving slowly downward at the surface of the gradient.

13. SDS and K⁺ form a precipitate at room temperature. Use samples right after boiling, or when using stored samples warm them up before loading. Avoid prolonged exposure to warm temperatures before loading.
14. Most proteins transfer well in the Western blot transfer buffer described in **Subheading 2.4.6.** Transfer conditions may need to be adjusted for your protein of interest. PVDF membranes are better for less abundant proteins and for stripping of blots. Please note that some PVDF membranes have high autofluorescence when used with the Odyssey Imagine System. Use Immobilon-FL from Milipore to avoid problems with autofluorescence with the Odyssey Imagine System.

Acknowledgments

This work was supported by a Walter V. and Idun Y. Berry Fellowship (R.V.), Deutsche Forschungsgemeinschaft Fellowship VO 864/1-1 (R.V.), a junior investigator award from the Stanford Digestive Disease Center (DDC DK56339; R.V., W.J.N.), and NIH RO1GM35227 (W.J.N.).

References

1. Knust, E. and Bossinger, O. (2002) Composition and formation of intercellular junctions in epithelial cells. *Science* **298,** 1955–1959.
2. Vogelmann, R., Amieva, M. R., Falkow, S., and Nelson, W. J. (2004) Breaking into the epithelial apical-junctional complex—news from pathogen hackers. *Curr. Opin. Cell Biol.* **16,** 86–93.
3. Nelson, W. J. (2003) Adaptation of core mechanisms to generate cell polarity. *Nature* **422,** 766–774.
4. Matter, K. and Balda, M. S. (2003) Signalling to and from tight junctions. *Nat. Rev. Mol. Cell Biol.* **4,** 225–236.
5. Ebnet, K., Suzuki, A., Ohno, S., and Vestweber, D. (2004) Junctional adhesion molecules (JAMs): more molecules with dual functions? *J. Cell Sci.* **117,** 19–29.
6. Grindstaff, K. K., Yeaman, C., Anandasabapathy, N., et al. (1998) Sec6/8 complex is recruited to cell–cell contacts and specifies transport vesicle delivery to the basal-lateral membrane in epithelial cells. *Cell* **93,** 731–740.
7. Yeaman, C. (2003) Ultracentrifugation-based approaches to study regulation of Sec6/8 (exocyst) complex function during development of epithelial cell polarity. *Methods* **30,** 198–206.
8. Yeaman, C., Grindstaff, K. K., Wright, J. R., and Nelson, W. J. (2001) Sec6/8 complexes on trans-Golgi network and plasma membrane regulate late stages of exocytosis in mammalian cells. *J. Cell Biol.* **55,** 593–604.
9. Vogelmann, R. and Nelson, W. J. (2005) Fractionation of the epithelial apical junctional complex: reassessment of protein distributions in different substructures. *Mol. Biol. Cell* **16,** 701–716.

3

Analysis of Neutrophil Chemotaxis

Paul A. Nuzzi, Mary A. Lokuta, and Anna Huttenlocher

Summary

Neutrophils are the initial responders to bacterial infection or other inflammatory stimuli and comprise a key component of the innate immune response. In addition to their unique morphology and antimicrobial activity, neutrophils are characterized by the ability to migrate rapidly up shallow gradients of attractants in vivo. The directed migration of neutrophils, referred to as chemotaxis, requires the temporal and spatial regulation of intracellular signaling pathways allowing the neutrophil to detect a gradient of attractant, polarize, and migrate rapidly toward the highest concentration of the chemoattractant. A challenge to understanding neutrophil chemotaxis is the inherent difficulty encountered when working with primary neutrophils, which are difficult to purify in the resting state, are not easily transfected, are terminally differentiated, and have a short life span after purification. Here we discuss neutrophil purification methods and chemotaxis assays and provide methodology for working with a neutrophil-like cell line, the HL-60 promyelocytic leukemia cell line. We also discuss methods for HL-60 transfection using retroviral approaches and chemotaxis assays used with differentiated HL-60 cells.

Key Words: Chemotaxis; neutrophil; HL-60 cell line; time-lapse video microscopy; Transwell assay.

1. Introduction

Neutrophil polarization and chemotaxis are the initial steps in the innate immune response. Neutrophils must respond to both endogenous and exogenous attractants to allow their recruitment to inflammatory sites, including sites of infection or wounding (1). Upon the inciting event, attractants are released that induce rapid changes in neutrophil shape, resulting in cell polarization and allowing the cell to orient and migrate up a gradient of attractant to the site of

From: *Methods in Molecular Biology, vol. 370:*
Adhesion Protein Protocols, Second Edition
Edited by: A. S. Coutts © Humana Press Inc., Totowa, NJ

inflammation *(2,3)*. This highly specialized form of migration requires both spatial and temporal regulation of the cytoskeleton *(4–6)*, numerous intracellular signaling proteins *(7–13)*, and membrane lipid reorganization *(14–16)*. As a result, neutrophils must integrate multiple exogenous signals and rapidly change direction in narrow gradients of attractants. However, the mechanisms that allow a cell to detect a gradient of attractant, prioritize divergent signals, and migrate directionally toward the attractant remain poorly understood. Many different methods have been developed to study chemotaxis of neutrophils and other cell types, including Transwell assay *(17)*, Dunn chemotaxis chamber *(18)*, under-agarose assay *(19)*, and time-lapse video microscopy toward a point source of attractant released from a pipet tip *(10–12,20)*. Here we describe detailed methods for primary neutrophil purification and the analysis of chemotaxis using Transwell assay and time-lapse microscopy.

A limitation of working with primary neutrophils is their short life span after purification (6–8 h), which makes it difficult to manipulate specific signaling pathways other than using inhibitor-based approaches. To circumvent these problems, other methods have been used to modify specific signaling pathways in primary neutrophils including HIV-TAT peptides *(10,21)* or the Semliki Forest viral system *(22)*. However, because these methods are technically difficult, their general utility is limited. A more feasible approach to dissect the contribution of specific signaling pathways to neutrophil chemotaxis involves the study of bone-marrow-derived neutrophils isolated from mice that have been genetically modified in a specific signaling pathway *(23,24)*. It is important to consider, however, that these are not mature neutrophils and that the purity of these preparations is less than that achieved with human neutrophil preparations (personal observation). However, again these types of studies do not allow for the efficient dissection of pathways that mediate neutrophil chemotaxis. An alternative approach used by many investigators is to utilize a neutrophil-like cell line. Since the isolation of the cell line in 1978 *(25)*, HL-60s have been used widely as a model system to study neutrophil chemotaxis *(20)*. HL-60s are a promyelocytic leukemia cell line that can be stably transfected, maintained in culture, and terminally differentiated (dHL-60) into a neutrophil-like cell using either dimethylsulfoxide (DMSO) or retinoic acid. Previous studies have shown that upon differentiation, dHL-60s display activation and cell-signaling phenotypes similar to primary neutrophils and respond to several chemoattractants such as C5a, fMLP, and LTB_4 *(26)*. HL-60 cell lines have also been used to visualize green fluorescent protein (GFP)-tagged proteins during chemotaxis providing a powerful tool to examine the temporal and spatial distribution of key signaling proteins during chemotaxis. Here we describe methods for HL-60 transfection using retroviral approaches and compare the methods to study chemotaxis of primary neutrophils and dHL-60 cells.

2. Materials

2.1. HL-60 Experiments

1. HL-60 promyelocytic cells (ATCC ref. no. CCL-240) (can also be obtained from UCSF Tissue Culture Facility).
2. Iscove's modified Dulbecco's medium + 10% heat-inactivated fetal bovine serum (FBS) and 1% penicillin-streptomycin.
3. 75-cm^2 Cell culture flask (T75).
4. Hemocytometer.
5. DMSO (Hybri-Max) (Sigma; cat. no. D2650).
6. Sterile Femtotips® micropipet tips (Eppendorf, cat. no. 930000035).
7. Eppendorf microloader pipet tips (Eppendorf, cat. no. 930001007).
8. 35 × 10 mm Non-tissue culture plastic Petri dish with an 18-mm-diameter hole cut in the center of the dish.
9. Circular glass cover slips (22 mm).
10. Norland Optical Adhesive 68 (Norland Products, New Brunswick, NJ).
11. Gey's medium *(26)*: 138 mM NaCl, 6 mM KCl, 1 mM Na$_2$HPO$_4$, 5 mM glucose, 20 mM N-hydroxyethylpiperazine-N'-2-ethanesulfonic acid (HEPES), 1 mM CaCl$_2$, 1 mM MgSO$_4$, and 0.5% human serum albumin (American Red Cross NDC 52769-251-05).
12. Eppendorf FemtoJet microinjection unit (Brinkmann Instruments, Westbury, NY).
13. Inverted microscope with a 60X differential interference contrast (DIC) oil objective, image acquisition system for time-lapse series, and closed system to maintain temperature at 37°C.

2.2. Generation of HL-60 Cell Lines

1. Retroviral expression vector with an IRES GFP element (LZRS-IRES or pMX-IRES vectors work best).
2. Amphotropic retroviral packaging cell line (e.g., Phoenix cell line, Nolan Lab, Stanford, CA).
3. Packaging cell growth medium: Dulbecco's modified Eagle's medium containing 10% heat-inactivated FBS and 1% penicillin-streptomycin.
4. 2X HBS: 50 mM HEPES, 10 mM KCl, 12 mM dextrose, 1.5 mM Na$_2$HPO$_4$ (dibasic, heptahydrate), 280 mM NaCl, brought to pH 7.0 using 1 N NaOH and kept at −20°C.
5. 4 mg/mL Polybrene® in ddH$_2$O and stored at −20°C.
6. 2 M CaCl$_2$ in ddH$_2$O and stored at −20°C.
7. 25 mM Chloroquine in ddH$_2$O and stored at −20°C.
8. 6-mm tissue culture-treated cell culture dishes.
9. 6-Well cell culture cluster plate (Costar, cat. no. 3516).
10. Water-jacketed tissue culture incubator set to 32°C and 5% CO$_2$.
11. Tabletop centrifuge with multiwell plate adaptors.

2.3. Primary Neutrophil Experiments

1. Polymorphprep™ (Nycomed, Oslo, Norway).

2. 1X Dulbecco's phosphate-buffered saline (PBS) without Ca^{2+} and Mg^{2+}.
3. 5X PBS without Ca^{2+} and Mg^{2+} made up in SiH_2O (Sigma, cat. no. P-3813).
4. Sterile irrigation water (SiH_2O; Baxter Healthcare Corporation, Deerfield, IL).
5. 5-mL Polypropylene round-bottom tubes (Falcon, cat. no. 2063).
6. USP preservative free sodium heparin at 1000 IU/mL (American Pharmaceutical Partners Inc.; Los Angeles, CA; NDC 63323-276-02).
7. 15-mL 17×120 mm Screw-cap conical tubes (Sarstedt).
8. 50-mL 17×120 mm Screw-cap conical tubes (Sarstedt).
9. Recombinant interleukin (IL)-8 (Sigma, cat. no. I-1645).
10. *N*-Formyl-Met-Leu-Phe (fMLP; Sigma, cat. no. F3506).
11. Recombinant complement factor 5a (C5a; Sigma, cat. no. C5788).
12. Fibrinogen (Sigma cat. no. F4883).
13. Fibronectin, purified according to Ruoslahti *(27)*.
14. 3-μm-Pore, 12-mm-diameter Transwells (Costar, cat. no. 3402).
15. EGM-2MV Medium (Cambrex Bio Science, Walkersville, MD, cat. no. CC-3202) (*see* **Note 1**).
16. Sterile 0.5 *M* ethylene diamine tetraacetic acid (EDTA).
17. Inverted microscope, 20X DIC objective, image acquisition system for time-lapse series and closed system to maintain temperature at 37°C (The Box closed system, Life Imaging Services).

3. Methods

In this section methods will be described for purification of primary neutrophils (**Subheading 3.1.**) and the growth, maintenance, and generation of HL-60 cell lines (**Subheadings 3.2.** and **3.3.**). Methodology is also provided for the Transwell assay (**Subheading 3.4.**) and micropipet time-lapse chemotaxis assay with both primary neutrophils (**Subheading 3.5.**) and dHL-60 cells (**Subheading 3.6.**).

3.1. Preparation of Primary Neutrophils

When working with human neutrophils, a detailed human subjects protocol must be established prior to starting the experiment. Special steps must also be taken to prevent the cells from becoming activated during the purification (*see* **Note 2**):

1. Peripheral human venous blood from normal healthy donors is drawn into a syringe (glass tubes will activate the cells) containing preservative-free heparin to a final concentration of 100 IU/mL, and the blood is used within 2 h of drawing.
2. Polymorphprep should be maintained between 18 and 22°C prior to the purification. Five to 7.5 mL of blood should be gently placed over an equal volume of Polymorphprep solution in a 15-mL screw-cap conical tube.

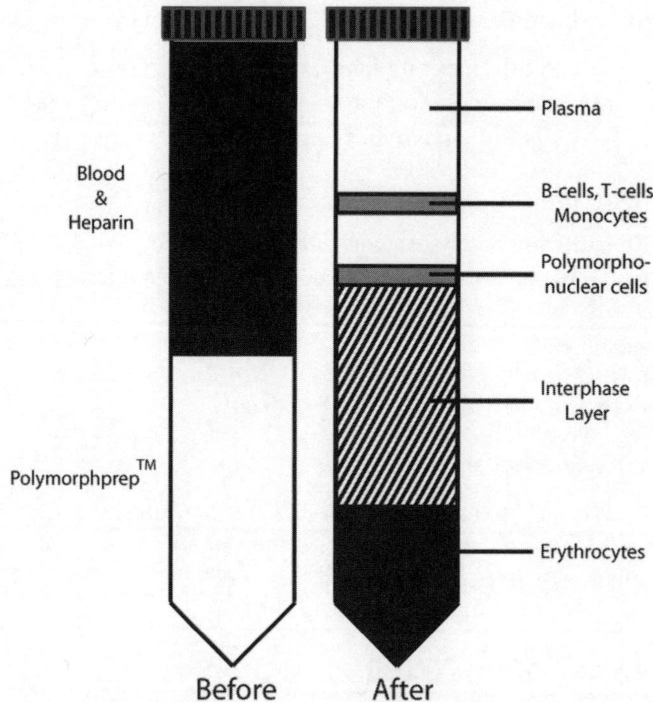

Before and after centrifugation

Fig. 1. Separation of human blood following centrifugation. Freshly isolated human blood is carefully placed over an equal volume of Polymorphprep solution. Following centrifugation, the blood is fractionated into several layers. The polymorphonuclear cells partition directly above the interphase layer and following further processing can be purified to ≥90% purity. (Adapted from the Polymorphprep instructions.)

3. The samples are then centrifuged according to the manufacturer's instructions at 18–22°C. Following centrifugation, two leukocyte bands should be visible (*see* **Fig. 1**).
4. Carefully aspirate off plasma and monocytic layer to prevent contamination.
5. Remove the neutrophil layer and combine cells from up to six tubes into 30 mL of 1X PBS, which is brought to 50 mL final volume and centrifuged at $400g$ for 10 min to remove residual Polymorphprep solution.
6. Decant supernatant.
7. To remove the contaminating red blood cells, resuspend pellet in 8 mL of SiH_2O for 30 s and immediately add 2 mL of 5X PBS.
8. Centrifuge at $228g$ for 5 min.
9. Remove supernatant and resuspend neutrophils in 1X PBS at a concentration ≤5.0 × 10^6 and store at 4°C to prevent the cells from clumping and activating.

3.2. Growth and Maintenance of HL-60 Cells

To generate stable cell lines and facilitate proper differentiation, HL-60 cells must be carefully maintained and passaged within a midlog range of growth. This requires the cells to be passaged approximately every 2 d.

3.2.1. Growth and Passage

1. Using a hemocytometer, count the HL-60 cells.
2. Resuspend the HL-60 cells at 2×10^5 cells/mL in 20–30 mL HL-60 growth medium and place them in a T75 tissue culture flask.
3. Cells are maintained at 37°C and 5% CO_2.
4. The cells are grown to a density of $\leq 1 \times 10^6$ cells/mL, at which time they are passaged by seeding a new flask at 2×10^5 cells/mL.

3.2.2. Freezing HL-60 Cells

1. Start with cells that are midlog growth (0.6–1×10^6 cells/mL).
2. Freeze 1×10^7 cells in regular growth medium supplemented with 10% DMSO.
3. Freeze cells slowly overnight at –80°C.
4. The next day, transfer to liquid nitrogen.

3.2.3. Cell Differentiation

1. Add 190 µL of 100% DMSO (hybridoma tested for lipopolysaccharide) to 15 mL of HL-60 cell medium.
2. Use this to resuspend 1×10^6 midlog growth HL-60 cells and place in a T75 tissue culture flask.
3. The cells are maintained at 37°C and 5% CO_2.
4. The cells will be differentiated and ready for use on days 6, 7, and 8 after addition of DMSO (*see* **Note 3**).

3.3. Generation of Stable HL-60 Cell Lines

To generate stable cell lines, retroviral-mediated approaches are used. Transfection efficiency can be as high as 30%. In brief, the gene of interest is inserted into the retrovirus expression vector and transfected transiently into the viral packaging cell line (Phoenix cells) using calcium phosphate precipitation. The viral supernatant is collected, and a spin infection of HL-60s is performed using the viral supernatant. Finally, fluorescence-activated cell sorting is performed for GFP 1 wk after infection to generate the cell lines.

1. One day prior to transfection, seed two 6-cm TC dishes with 1.5×10^6 packaging cells in 3 mL of packaging cell medium for each line you wish to make (*see* **Note 4**).
2. The following day, each plate should be approx 60% confluent.
3. Ten minutes prior to transfection, 4 µL chloroquine is added directly to each plate and swirled to mix.

4. For each 6-cm dish, the transfection mixture is created as follows:
   a. Tube 1: 12 µg DNA vector, 62 µL of 2 M $CaCl_2$, and scale to 500 µL with ddH_2O.
   b. Tube 2: 500 µL of warm 2X HBS buffer.
5. Tube 1 is then carefully bubbled into Tube 2.
6. The mixture is then added dropwise to one 6-cm dish of cells and swirled to mix.
7. The cells are then incubated for 6 h at 37°C and 5% CO_2.
8. Following the 6-h incubation, each plate is gently washed with PBS, and 3 mL of fresh packaging cell medium added.
9. At 24 h posttransfection, the cells are moved to a different incubator and maintained at 32°C and 5% CO_2 (*see* **Note 5**).
10. At 48 h posttransfection, the medium is removed from one of the 6-cm plates and filtered through a 0.45-µm syringe filter into a conical tube.
11. Polybrene® is added to the supernatant to a final concentration of 4 µg/mL, and the viral supernatant/polybrene mixture is used to resuspend 1×10^6 midlog growth HL-60s.
12. The cells are then placed in one well of a six-well cell culture cluster plate and centrifuged at 1141g at 32°C for 90 min in a tabletop centrifuge (*see* **Note 6**).
13. Following the spin, the cells are put back at 32°C and allowed to grow for 6 h.
14. The cells are then pelleted down, resuspended in HL-60 growth medium at 0.5×10^6/mL and transferred back to 37°C (*see* **Note 7**).
15. The following day, an additional infection of the same cells is carried out (repeat **steps 10–14**) using the second transfected 6-cm dish of packaging cells and the previously infected HL-60s.
16. At 48–72 h postinfection, transfection efficiency can be assayed by visualization of the GFP.
17. The cells should then be passaged and processed for fluorescence-activated cell sorting to generate stable cell lines.

3.4. Transwell Assay for Neutrophils and dHL-60s

In this section we describe modifications of a Boyden Chamber assay or Transwell assay for the study of neutrophil chemotaxis. The transwell assay can be used to examine random cell motility or chemokinesis with soluble attractant in both the upper and lower compartments of the Transwell chamber or, alternatively, to examine chemotaxis or directed migration toward higher concentrations of soluble attractant in the lower compartment of the chamber only. The functional readout of the assay is the number of cells that migrate to the bottom chamber in a specific time period and does not provide information about the migration phenotype that may contribute to altered chemotaxis. However, the advantage of the Transwell assay is the ability to perform replicates and to compare chemotaxis under different conditions (dose–response effects with different concentrations or types of attractants or extracellular matrix components).

1. The night before the assay, coat two 3-µm-pore, 12-mm-diameter Transwells for each experimental condition with 2.5 µg/mL fibrinogen. To coat, add 600 µL in the

bottom well and 200 µL on top of the membrane and incubate at 37°C for 1 h. Other extracellular matrix proteins can also be used to coat the membrane.
2. The Transwells are then washed two times with PBS.
3. Carefully aspirate off the residual PBS and allow the Transwells to dry overnight in a laminar flow hood (*see* **Note 8**).
4. Use 8×10^5 cells for each experimental condition, and resuspend the cells to 400 µL in medium.
5. Add 600 µL of medium containing chemoattractant to the bottom of each well and 200 µL of this cell preparation to the top of each of two filters (for replicates) (*see* **Note 9**).
6. For chemoattractants we use 1.25 nM IL-8, 100 nM fMLP, or 11 nM C5a. IL-8, fMLP, and C5a are all very effective for primary neutrophil chemotaxis, while C5a is most effective for the dHL-60s.
7. The cells are allowed to migrate for 2 h at 37°C and 10% CO_2.
8. Following incubation, 60 µL of 0.5 M EDTA is added to the bottom chamber, and the plate is incubated 10–15 min at 4°C.
9. The filters are then removed from the well, and the number of cells in the bottom of each well is counted using the hemocytometer.

3.5. Time-Lapse Chemotaxis Assay for Primary Neutrophils

Time-lapse microscopy of neutrophils migrating to a micropipet tip that releases attractant allows the tracking of cell speed and persistence time and provides information about the mechanism for altered migration *(10)*. The source of attractant can also be moved to change the direction of the attractant source. Chemotaxis assays are generally performed using 20X or 40X DIC objective, but if immunofluorescence of GFP-tagged proteins is being examined, 60X or 100X DIC objectives are recommended.

1. Place 3 mL of EGM-2MV medium in a loosely capped 5-mL polypropylene tube kept at 37°C and 10% CO_2 to charge the medium with CO_2 and keep the medium warm prior to the experiment.
2. Coat a 35-mm non-TC-treated dish with 2 mL of 2.5 µg/mL fibrinogen in PBS at 37°C for 1 h and then rinse two times with PBS.
3. 5×10^5 neutrophils are added to the 3 mL of EGM-2MV, incubated for 20 min, plated, and allowed to settle for 10 min prior to performing the experiment. For inhibitor studies the neutrophils are pretreated with the inhibitor at specified concentrations accordingly.
4. Chemoattractant is centrifuged for 5 min at 16,000g to remove particulate matter. Three µL of chemoattractant (12.5 µM IL-8, 100 µM fMLP, or 85 µM C5a) is loaded into a micropipet using a microloader pipet tip (*see* **Note 10**).
5. The micropipet is screwed in place, the microinjection unit turned on, and the back-pressure (P_c) is set to 10 hPa.

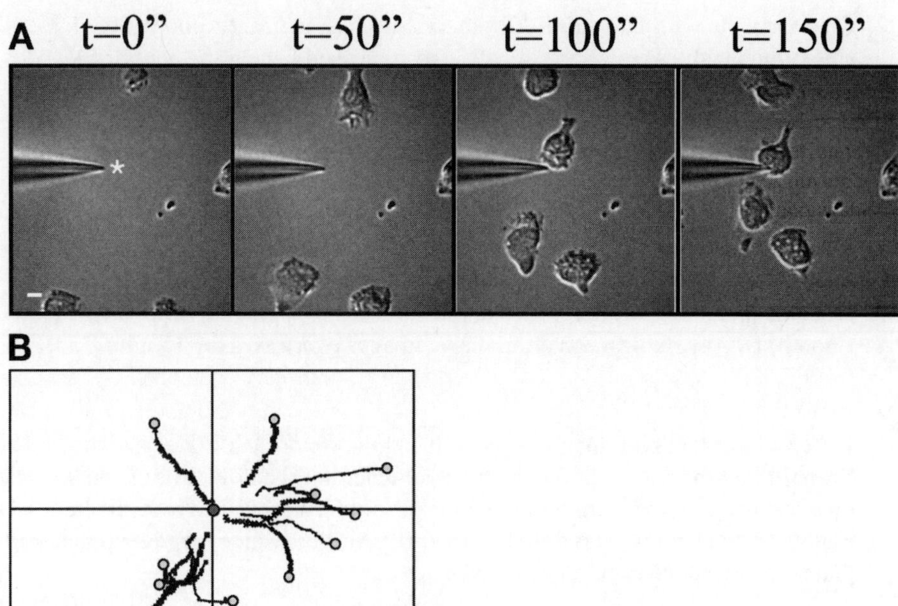

Fig. 2. Time-lapse microscopy of differentiated HL-60 cells to C5a. **(A)** dHL-60 cells were plated onto dishes coated with 2.5 μg/mL fibrinogen and 100 μg/mL fibronectin, and a chemotactic gradient was generated using a micropipet tip loaded with C5a. Images were collected at 10-s intervals for 10 min. Images are shown at time points t = 0, 50, 100, and 150 s. The tip of the micropipet is marked with an asterisk. Bar = 5 μm. **(B)** The rose plot shows the path of each cell tracked relative to the micropipet tip designated (0,0).

6. The plate is then placed on the microscope and the micropipet lowered and brought into focus (*see* **Note 11**).
7. Time-lapse images are then captured every 15 s for 20 min.

3.6. Time-Lapse Chemotaxis Assay for dHL-60 Cells

The chemotaxis of dHL-60 cell lines that overexpress specific proteins or express dominant-negative forms of proteins can be used to determine the role of specific signaling pathways during chemotaxis (*see* **Fig. 2**). This assay provides information about the morphology and migration defects in established HL-60 cell lines. The assay can also be used analyze the temporal and spatial distribution of specific GFP-tagged proteins during chemotaxis. We use a 60X or 100X DIC oil objective to provide better resolution.

1. Prepare 1 glass-bottom plate for each experimental condition (*see* **Note 12**).

2. Coat the plate with 100 µg/mL fibronectin and 2.5 µg/mL fibrinogen for 1 h, and rinse twice with PBS.
3. Resuspend 5×10^5 dHL-60s in 3 mL prewarmed, CO_2-charged Gey's medium and plate for 10 min at 37°C and 5% CO_2.
4. Chemoattractant is centrifuged for 5 min at 16,000g to remove particulate matter. Load 3 µL of chemoattractant (85 µM C5a) into a micropipet using a microloader pipet tip (*see* **Note 10**).
5. Screw the micropipet in place: the microinjection unit and the P_c is set to 12 hPa (*see* **Note 11**).
6. Place the plate on the microscope and lower the micropipet to bring it into focus.
7. Capture DIC or fluorescence time-lapse images every 10 s for 15 min.

4. Notes

1. For experiments involving primary neutrophils, we use the EGM-2MV medium. Primary neutrophils do not display an activated morphology in this media. The medium is a multipart mixture purchased from Cambrex. The base medium is a bottle of Endothelial Cell Basal Medium-2 (EBM-2), to which several supplied supplements are added to produce EGM-2MV.
2. A word of caution: working with primary neutrophils can be very difficult, and it may take several attempts to become proficient at purification without activating the neutrophils. During the procedure, great care should be taken to ensure that the centrifuge is properly balanced, because minor vibrations can activate the cells during the purification. While carrying out all centrifugation steps, *the brake should not be applied*, because this can lead to cell activation. Neutrophils will also rapidly activate when exposed to any glass surface or when resuspended to a concentration greater than the suggested 5.0×10^6 cell/mL. Finally, all steps during the purification should be performed as quickly as possible. Allowing the cells to sit during many of the purification steps can result in neutrophil activation.
3. Upon differentiation, many of the dHL-60 cells will become polarized and activated when counted on a glass hemocytometer, similar to our observations with primary neutrophils. For experiments in which a large number of dHL-60s cells is required, the protocol can be doubled (380 µL DMSO, 2×10^6 cells, and 30 mL final volume of medium) and still be accommodated in a T75 flask.
4. It is critical that the retroviral packaging cell medium and the HL-60 cell medium are made with heat-inactivated FBS. Additionally, it is important to make a stable cell line expressing the empty vector as a negative control.
5. The retrovirus produced is much more stable when produced at 32°C.
6. For suspension cells, we find that centrifugation of the viral supernatant with the cells enhances the efficiency of the viral infection. We use an Allegra 6R tabletop centrifuge (Beckman Coulter) with multiwell plate adaptors to accommodate the six-well tissue culture plates. The centrifuge should also have a temperature setting to maintain the cells at 32°C during the spin.
7. Incubation with the viral supernatant can go slightly longer than 6 h, but prolonged exposure to Polybrene is toxic to the cells.

8. The Transwells must be completely dry prior to the start of the experiment. If they are still damp, they will leak and not establish a proper gradient.

9. We consistently use EGM-2MV for primary neutrophils and Gey's medium for dHL-60 cells. It is important to note that the medium in the top and bottom of the wells must be the same to establish a proper gradient.

10. Centrifugation of the chemoattractant is important prior to loading the needle because this assay is sensitive to false-negative results in that the needle can be easily clogged. Therefore, positive controls before and after the sample are required.

11. Because we are not injecting the cells, the values for the injection pressure (P_i) and injection time (T_i) are irrelevant when setting up the microinjection unit. With a backpressure of 10 hPa for neutrophils and 12 hPa for dHL-60s, chemoattractant is being released into the medium while the micropipet is being brought into focus. Therefore, after all the minor adjustments are made, we move the micropipet to a new field and begin the movies. This helps to ensure that the movie starts as the gradient is first being established. *The height and the angle of the micropipet in relation to the bottom of the dish is critical for proper gradient formation and, therefore, chemotaxis.* Because all microscope setups are different, it is best to determine this empirically. Once optimized, we created a template with the proper angle to shorten setup time and standardize each experiment.

12. For these experiments a 35-mm dish is created with a glass bottom to allow for the 60X oil emersion objective and fluorescence microscopy. To make each plate, start with a 35 × 10 mm non-TC plastic Petri dish that has an 18-mm-diameter hole cut in the middle of the dish. (We make a large stock of these in the lab using a drill press.) On the underside of the plate, place a small amount of optical adhesive around the periphery of the hole. This adhesive will not begin to cure until exposed to long-wave UV light (~350 nm). Place a 22-mm circular glass cover slip onto the glue to close the opening, and incubate under a UV light source on each side prior to use. The exposure time required to completely cure the glue is dependent on both the intensity of the source and its height from the plate. For additional details, *see* the Optical Adhesive 68 product sheet or the Norland website (https://www.norlandprod.com).

Acknowledgments

The authors would like to thank researchers in the Huttenlocher laboratory for useful discussions. We would like to thank Kate Cooper for critical reading of the manuscript. In addition, we would like to acknowledge colleagues in the field, notably Henry Bourne and members of his laboratory, who provided guidance with the HL-60 cells. This work was supported by the American Heart Association O255769N, 0225401Z and National Institutes of Health R01 GM074827.

References

1. Ley, K. (2002) Integration of inflammatory signals by rolling neutrophils. *Immunol. Rev.* **186,** 8–18.

 2. Parent, C. A. and Devreotes, P. N. (1999) A cell's sense of direction. *Science* **284,** 65–70.
 3. Bourne, H. R. and Weiner, O. (2002) A chemical compass. *Nature* **419,** 21.
 4. Weiner, O. D., Servant, G., Welch, M. D., Mitchison, T. J., Sedat, J. W., and Bourne, H. R. (1999) Spatial control of actin polymerization during neutrophil chemotaxis. *Nat. Cell Biol.* **1,** 75–81.
 5. Niggli, V. (1999) Rho-kinase in human neutrophils: a role in signalling for myosin light chain phosphorylation and cell migration. *FEBS Lett.* **445,** 69–72.
 6. Eddy, R. J., Pierini, L. M., Matsumura, F., and Maxfield, F. R. (2000) Ca^{2+}-dependent myosin II activation is required for uropod retraction during neutrophil migration. *J. Cell Sci.* **113 (Pt. 7),** 1287–1298.
 7. Funamoto, S., Meili, R., Lee, S., Parry, L., and Firtel, R. A. (2002) Spatial and temporal regulation of 3-phosphoinositides by PI 3-kinase and PTEN mediates chemotaxis. *Cell* **109,** 611–623.
 8. Haugh, J. M., Codazzi, F., Teruel, M., and Meyer, T. (2000) Spatial sensing in fibroblasts mediated by 3' phosphoinositides. *J. Cell Biol.* **151,** 1269–1280.
 9. Kraynov, V. S., Chamberlain, C., Bokoch, G. M., Schwartz, M. A., Slabaugh, S., and Hahn, K. M. (2000) Localized Rac activation dynamics visualized in living cells. *Science* **290,** 333–337.
10. Lokuta, M. A., Nuzzi, P. A., and Huttenlocher, A. (2003) Calpain regulates neutrophil chemotaxis. *Proc. Natl. Acad. Sci. USA* **100,** 4006–4011.
11. Servant, G., Weiner, O. D., Herzmark, P., Balla, T., Sedat, J. W., and Bourne, H. R. (2000) Polarization of chemoattractant receptor signaling during neutrophil chemotaxis. *Science* **287,** 1037–1040.
12. Wang, F., Herzmark, P., Weiner, O. D., Srinivasan, S., Servant, G., and Bourne, H. R. (2002) Lipid products of PI(3)Ks maintain persistent cell polarity and directed motility in neutrophils. *Nat. Cell Biol.* **4,** 513–518.
13. Weiner, O. D., Neilsen, P. O., Prestwich, G. D., Kirschner, M. W., Cantley, L. C., and Bourne, H. R. (2002) A PtdInsP(3)- and Rho GTPase-mediated positive feedback loop regulates neutrophil polarity. *Nat. Cell Biol.* **4,** 509–513.
14. Seveau, S., Eddy, R. J., Maxfield, F. R., and Pierini, L. M. (2001) Cytoskeleton-dependent membrane domain segregation during neutrophil polarization. *Mol. Biol. Cell* **12,** 3550–3562.
15. Pierini, L. M., Eddy, R. J., Fuortes, M., Seveau, S., Casulo, C., and Maxfield, F. R. (2003) Membrane lipid organization is critical for human neutrophil polarization. *J. Biol. Chem.* **278,** 10,831–10,841.
16. Manes, S., Ana Lacalle, R., Gomez-Mouton, C., and Martinez, A. C. (2003) From rafts to crafts: membrane asymmetry in moving cells. *Trends Immunol.* **24,** 320–326.
17. Huttenlocher, A., Ginsberg, M. H., and Horwitz, A. F. (1996) Modulation of cell migration by integrin-mediated cytoskeletal linkages and ligand-binding affinity. *J. Cell Biol.* **134,** 1551–1562.
18. Wells, C. M. and Ridley, A. J. (2005) Analysis of cell migration using the Dunn chemotaxis chamber and time-lapse microscopy. *Methods Mol. Biol.* **294,** 31–41.

19. Heit, B. and Kubes, P. (2003) Measuring chemotaxis and chemokinesis: the under-agarose cell migration assay. *Sci. STKE* **2003,** PL5.
20. Servant, G., Weiner, O. D., Neptune, E. R., Sedat, J. W., and Bourne, H. R. (1999) Dynamics of a chemoattractant receptor in living neutrophils during chemotaxis. *Mol. Biol. Cell* **10,** 1163–1178.
21. Alblas, J., Ulfman, L., Hordijk, P., and Koenderman, L. (2001) Activation of Rhoa and ROCK are essential for detachment of migrating leukocytes. *Mol. Biol. Cell* **12,** 2137–2145.
22. Stofega, M. R., Sanders, L. C., Gardiner, E. M., and Bokoch, G. M. (2004) Constitutive p21-activated kinase (PAK) activation in breast cancer cells as a result of mislocalization of PAK to focal adhesions. *Mol. Biol. Cell* **15,** 2965–2977.
23. Li, S., Yamauchi, A., Marchal, C. C., Molitoris, J. K., Quilliam, L. A., and Dinauer, M. C. (2002) Chemoattractant-stimulated Rac activation in wild-type and Rac2-deficient murine neutrophils: preferential activation of Rac2 and Rac2 gene dosage effect on neutrophil functions. *J. Immunol.* **169,** 5043–5051.
24. Glogauer, M., Marchal, C. C., Zhu, F., et al. (2003) Rac1 deletion in mouse neutrophils has selective effects on neutrophil functions. *J. Immunol.* **170,** 5652–5657.
25. Collins, S. J., Ruscetti, F. W., Gallagher, R. E., and Gallo, R. C. (1978) Terminal differentiation of human promyelocytic leukemia cells induced by dimethyl sulfoxide and other polar compounds. *Proc. Natl. Acad. Sci. USA* **75,** 2458–2462.
26. Hauert, A. B., Martinelli, S., Marone, C., and Niggli, V. (2002) Differentiated HL-60 cells are a valid model system for the analysis of human neutrophil migration and chemotaxis. *Int. J. Biochem. Cell Biol.* **34,** 838–854.
27. Ruoslahti, E., Hayman, E. G., Pierschbacher, M., and Engvall, E. (1982) Fibronectin: purification, immunochemical properties, and biological activities. *Methods Enzymol.* **82 (Pt. A),** 803–831.

4

Analysis of Leukocyte Migration Through Monolayers of Cultured Endothelial Cells

Helen M. McGettrick, Lynn M. Butler, and Gerard B. Nash

Summary

In this chapter methods are described for analyzing the adhesion and migration of isolated leukocytes on endothelial cell monolayers that have been cultured on different substrates and treated with cytokines. When endothelial cells are grown on porous filters inserted in wells, the levels of leukocyte adhesion and migration are calculated from the number added and the numbers retrieved from the upper and lower chambers. Fluorescence microscopic examination of the fixed filters can be used to ascertain whether leukocytes are retained above or below the filter. Direct observations of the time course of migration can be made when endothelial cells are cultured in six-well plates after leukocytes are allowed to settle onto them for a short period. In a more specialized assay, leukocytes are perfused through glass capillaries coated with endothelial cells, and again, direct video-microscopic observations are made. In this assay all stages of capture, immobilization, and migration can be followed. In general, the filter-based assay has the highest throughput and greatest ease of use but yields less detailed information, whereas the flow-based assay is most difficult to set up but is most physiologically relevant.

Key Words: Leukocyte; neutrophil; endothelial cells; adhesion; migration; cytokines; cell culture.

1. Introduction

All classes of leukocytes must move from the circulation into tissue to carry out their protective functions. To achieve the transfer, the flowing cells must adhere to vascular endothelial cells and migrate through the vessel wall (*see* **refs.** *1* and *2* for review of the stages in migration). In general, capture from flow is enabled through endothelial receptors (such as selectins) that have particularly rapid forward rate constants and are able to bind ligands presented by the fast-moving leukocytes. Because these receptors also have a rapid reverse

From: *Methods in Molecular Biology, vol. 370:*
Adhesion Protein Protocols, Second Edition
Edited by: A. S. Coutts © Humana Press Inc., Totowa, NJ

rate constant, an unstable rolling form of adhesion may follow, until the leukocytes receive an activating signal at the endothelial surface. Such signals (e.g., from chemokines specific for different leukocyte subsets) induce a conformational change in integrin receptors on the leukocytes, which then bind their cognate receptors on the endothelium (such as intercellular adhesion molecule-1). This binding is relatively stable and is able to anchor the leukocytes while they spread on the surface and migrate over it and then through the monolayer. The complete adhesion and migration response can be driven and controlled by the endothelial cells if they are exposed to inflammatory cytokines such as tumor necrosis factor (TNF)-α or interleukin (IL)-1β.

This chapter describes several different in vitro methods of analyzing the migration of leukocytes through endothelial monolayers in the presence or absence of flow. All of the methods yield information about the number of leukocytes adhering as well as the proportion of adherent cells that successfully transmigrate through the monolayer. The first method is based on analysis of the migration of leukocytes through monolayers cultured on porous Transwell filters. This type of method *(3–6)* has the advantages of not requiring special apparatus, having relatively high throughput, and being quite conservative in its requirements for numbers of endothelial cells and leukocytes per test. It is, however, something of a "black box." Typically the number of leukocytes passing through the filter and into a lower chamber is assessed. The location of other cells in the filter is not known (e.g., above or below the endothelial monolayer, and potentially adhered to the bottom of the filter). The kinetics of appearance of cells into the chamber below the filter is much slower than the kinetics of movement through an endothelial monolayer. Transmigration can be enhanced by adding a chemotactic agent to the lower chamber, but in such cases one is effectively testing chemotaxis with endothelial cells in the way, rather than endothelial-driven migration. The control systems cannot be expected to be the same in each case.

A direct visual assay of adhesion and migration through monolayers cultured on transparent plastic plates is also described (adapted from **ref. 7**). The rate of transmigration can be characterized, as well as the velocities of the migrating cells both on top of and underneath the monolayer. The method relies on cytokine treatment of the endothelial cells, which then effectively deliver the requirements for adhesion and migration. Finally, a flow-based assay is described in which leukocytes are perfused through glass capillaries, which are coated with endothelial cells *(8,9)*. Although this requires more specialized culture and adhesion apparatus, each stage in migration can be assessed (capture–activation–migration over, through, and under the monolayer), along with its kinetics. Moreover, several studies suggest that the presence of shear forces imposed by flow can modify the migratory behavior of the leukocytes *(10–12)*. Thus, this system may be the most physiologically relevant of the in vitro assays.

In principle, all of the assays can be used for any leukocyte subset. We have the most experience in studies of neutrophilic granulocytes (neutrophils), but have also used the assays for mixed mononuclear cells and peripheral blood lymphocytes (PBL). Others have used similar methods in studies of monocytes, preseparated lymphocyte subsets, and T-cell lines *(13–16)*. The endothelial cells described here are primary human umbilical vein endothelial cells (HUVEC), which are the most widely used and best characterized cells for this type of work. Other sources of primary cells and immortalized cells lines can be used; the only caveat is that requirements for stimulation and levels of adhesion and migration may vary between cell types. For instance, in our hands, when immortalized human dermal microvascular endothelial cells *(17)* are treated with TNF, they support transmigration of PBL but not neutrophils, whereas HUVEC support transmigration of both.

2. Materials

2.1. Blood Cell Isolation

1. K_2-Ethylene diamine tetraacetic acid (EDTA) in 10-mL tubes (Sarstedt, Numbrecht, Germany).
2. Histopaque 1077 (H1077) and Histopaque 1119 (H1119) (Sigma-Aldrich, Poole, UK).
3. Phosphate-buffered saline with 1 mM Ca^{2+} or 0.5 mM Mg^{2+} (PBS; Gibco, Invitrogen Ltd., Paisley, UK) with 0.15% (w/v) bovine serum albumin (BSA; dilute from 7.5% culture-tested solution; Sigma) and 5 mM glucose (PBSA).
4. M199-BSA: Medium 199 (M199; Gibco) supplemented with 0.15% (w/v) BSA (M199BSA).
5. Isotonic 2% glutaraldehyde in PBS (Cowley, Oxford, UK).
6. Nuclear stain, bisbenzamide (1 µg/mL; Sigma).

2.2. Culture of Endothelial Cells

1. M199 supplemented with gentamycin sulfate (35 µg/mL), human epidermal growth factor (10 ng/mL; Sigma E9644), and fetal calf serum (20% v/v heat-inactivated) (all from Sigma). Adding hydrocortisone (1 µg/mL, from 10 mg/mL stock in ethanol; Sigma) improves growth if going beyond first passage.
2. Bovine skin gelatin (Type B, 2% solution, culture tested; Sigma).
3. Collagenase (type IA; Sigma) stored at –20°C at 10 mg/mL in PBS, thawed and diluted to 1 mg/mL with M199 for use.
4. Autoclaved cannulae and plastic ties (electrical).
5. EDTA solution (0.02%, culture tested; Sigma).
6. 70% (v/v) Ethanol or industrial methylated spirits.
7. TNF-α and IL-1β (both from Sigma), stored in small aliquots at –80°C.

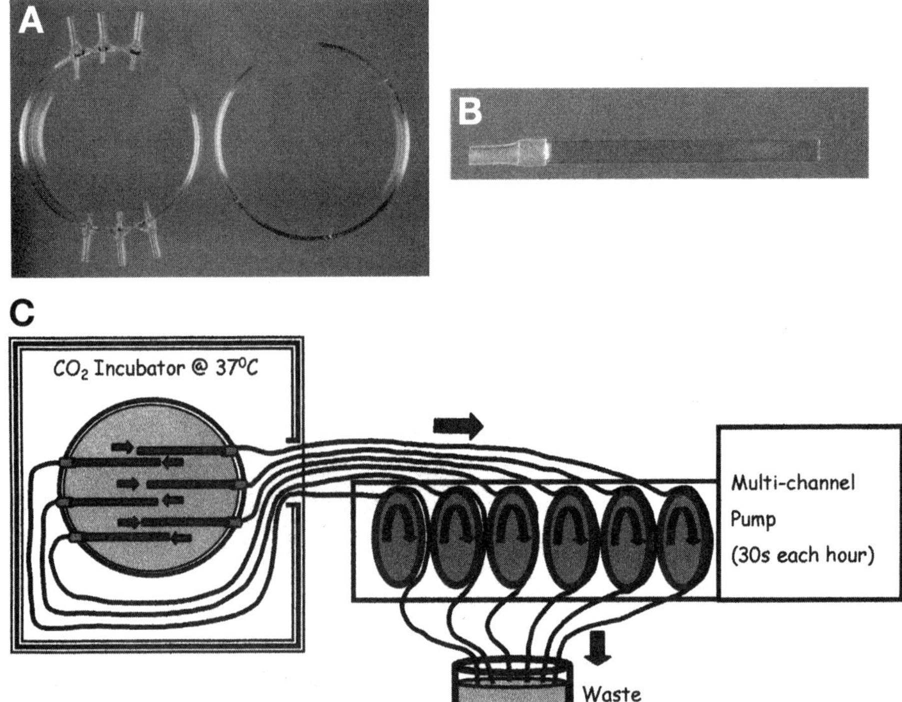

Fig. 1. Apparatus for culture of endothelial cells in microslides. (**A**) Glass culture dish with six ports fused into the wall; diameter 90 mm; volume of culture approx 50 mL. (**B**) Microslide with silicone rubber tubing adapter attached. (**C**) Schematic diagram of apparatus for culture of endothelial cells in six microslides. The glass culture dish with six microslides attached is placed in a CO_2 incubator. Tubing attached to each outlet passes through a port in the incubator wall. Perfusion is supplied for 30 s in each hour by a multichannel roller pump that pumps to waste.

2.3. Surfaces for Endothelial Culture

1. Transwell system: low-density 3.0-µm-pore polycarbonate Transwell filters (referred to as filters in future text) and matching filter plates (BD Pharmingen, Oxford, UK).
2. Six-well plates (BD Pharmingen).
3. Microslides: glass capillaries with rectangular cross section (0.3 mm × 3 mm; length 50 mm) (Vitro Dynamics Inc., Rockaway, NJ).
4. Aminopropyltriethoxysilane 4% (v/v) in acetone (Sigma) with molecular sieve (BDH Laboratory Supplies, Poole, UK) added to ensure anhydrous.
5. A special culture dish for use with microslides (**Fig. 1**): constructed by fusing six glass tubing side arms into the wall of a Pyrex glass Petri dish 100 mm in diameter (Fisons Scientific Equipment Ltd., Leicester, UK) (made to order by the Glassblowing Workshop of the School of Chemisty, The University of Birmingham, UK).

2.4. Flow-Based Adhesion Assay

1. Flow system: syringe pump with smooth flow (e.g., PHD2000 infusion/withdrawal, Harvard Apparatus, South Natick, MA). Electronic three-way microvalve with zero dead volume (LFYA1226032H Lee Products Ltd., Gerrards Cross, Buckinghamshire, UK) with 12 V DC power supply for valve. Silicon rubber tubing, internal diameter (ID)/external diameter of 1/3 mm and 2/4 mm (Fisher Scientific, Loughborough, UK). Scotch double-sided adhesive tape, approx 1 cm wide (3M Ltd, Bracknell, Berkshire, UK). Three-way stopcocks (BOC Ohmeda AB, Helsinborg, Sweden). Sterile, disposable syringes (2, 5, and 10 mL; Becton Dickinson, Oxford, UK) and glass 50-mL syringe for pump (Popper Micromate; Popper and Sons Inc., New York).
2. Video-microscope: microscope with heated stage or, preferably, with stage and attached flow apparatus enclosed in a temperature-controlled chamber at 37°C and phase contrast optics. Fluorescence capability is desirable for some variants of assay. Also useful are video camera (e.g., analog Cohu 4912 monochrome camera with remote gain control), monitor and video recorder (e.g., time-lapse, Panasonic AG-6730).
3. Image analysis: computer with video capture card and specialist software for counting cells, measuring motion, etc. A range of commercial packages is available, and image-analysis software (NIH Image http: //rsb.info.nih.gov/nih-image/) is available free over the Internet. We currently use Image Pro software (DataCell Limited, Finchampstead, UK).

3. Methods

3.1. Leukocyte Isolation (see Note 1)

1. Draw blood from the antecubital vein of normal human volunteers with a minimum of stasis, dispense into K_2-EDTA tubes, and mix gently but thoroughly.
2. Place 2.5 mL H1119 in 10-mL centrifuge tube and gently layer 2.5 mL H1077 on top. An interface should be visible between the layers.
3. Layer whole blood (5 mL) on top (**Fig. 2**).
4. Centrifuge at 800g for 30 min.
5. Retrieve the mononuclear cells from the top of the gradient at the interface of plasma and H1077 (**Fig. 2**).
6. Discard the middle section of H1077.
7. Retrieve the granulocyte (mainly neutrophil) layer from the H1077/H1119 interface.
8. Wash cells twice in PBSA with Ca^{2+}/Mg^{2+} or M199BSA.
9. To deplete mononuclear cells of monocytes, place in culture dish for 30 min at 37°C for monocytes to sediment and adhere. Gently wash off enriched lymphocytes. To obtain monocytes, a separate isolation procedure is advisable (*see* **Note 1**).
10. Count leukocytes and dilute to desired concentration in PBSA with Ca^{2+}/Mg^{2+} or M199BSA or endothelial culture medium (*see* **Note 2**).

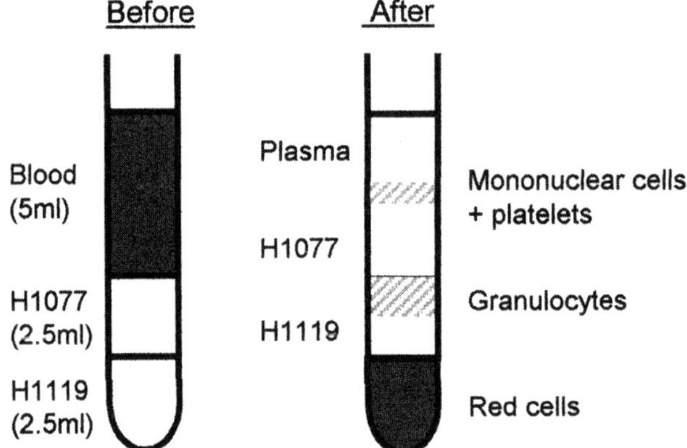

Fig. 2. Layering of liquids used for density gradient fractionation of blood. Before: 2.5 mL Histopaque 1077 (H1077) layered on 2.5 mL Histopaque 1119 (H1119), with 5 mL blood layered on top. After: redistributed cells and liquids after centrifugation at 800*g* for 30 min.

3.2. Culture of Endothelial Cells on Various Surfaces

There are various methods for culture of endothelial cells from different sources, and for the novice it is probably best to start by buying cells and media from commercial suppliers. Our current method for isolating and culturing human umbilical vein endothelial cells is given below, adapted from Cooke et al. *(18)*, which also described methods for coating microslides and culturing HUVEC in them (**Subheading 3.2.5.**).

3.2.1. Isolation and Primary Culture of HUVEC

1. Place the cord on paper toweling in a tray and spray liberally with the 70% ethanol. Choose sections of about 3–4 in. that do not have any clamp damage. Each 3- to 4-in. piece of cord equates to one flask of primary cells.
2. Locate the two arteries and one vein at one end of the cord.
3. Cannulate the vein and secure the cannula with an electrical tie.
4. Carefully wash through the vein with PBS using a syringe and blow air through to remove the PBS.
5. Cannulate the opposite end of the vein and tie off.
6. Inject collagenase (~10 mL per 3–4 in.) into vein until both cannulae bulbs have the mixture in them.
7. Place the cord into an incubator for 15 min at 37°C.
8. Remove from the incubator and tighten the ties. Massage the cord for approx 1 min.
9. Flush the cord through using a syringe and 10 mL PBS into a 50-mL centrifuge tube.
10. Push air through to remove any PBS; repeat this twice more (3 × 10 mL).
11. Centrifuge at 400*g* for 5 min. Discard supernatant.
12. Resuspend the cells in approx 1 mL of culture medium and mix well with pipet.

13. Make up to 4 mL in complete medium.
14. Add cell suspension to a 25-cm^2 culture flask.
15. Change medium after 2 h and again the next day. Cells should be confluent in about 3–7 d.

3.2.2. Dispersal of Endothelial Monolayers for Seeding New Surfaces

1. Rinse a flask containing a confluent primary monolayer of HUVEC with 2 mL of EDTA solution.
2. Add 2 mL of trypsin solution and 1 mL of EDTA for 1–2 min at room temperature, until the cells became detached. Tap on bench to loosen.
3. Add 8 mL of culture medium to the flask, remove the resulting suspension, and centrifuge at 400g for 5 min.
4. Remove supernatant and resuspend the cell pellet in 0.5 mL of culture medium and disperse by sucking them in and out of a pipet tip.
5. Make up to desired volume of culture medium for seeding onto different surfaces.

3.2.3. Seeding HUVEC on Transwell Filters

1. Add 700 µL of culture medium to each well in a 24-well plate (lower chamber).
2. Use sterile forceps to place an uncoated filter into each well (*see* **Note 3**).
3. Trypsinize a single flask of HUVEC as in **Subheading 3.2.2.**
4. Make up to 8 mL with culture medium and add 200 µL into the top of each filter (upper chamber) (*see* **Note 4**).
5. Replace the medium 24 h later and culture for 1–4 d (*see* **Note 5**).
6. Add TNF (100 U/mL) or IL-1 (5×10^{-11} g/mL) to upper and lower chambers for desired period before migration assay (typically 4 h for neutrophils or 24 h for PBL).

3.2.4. Seeding HUVEC in Six-Well Plates

1. Add 1 mL of 1% gelatin (in PBS) to each well for 15 min.
2. Trypsinize a single flask of HUVEC as in **Subheading 3.2.2.**
3. Make up to 8 mL with culture medium (*see* **Note 6**) and add 2 mL of HUVEC suspension to each of four wells.
4. Replace the medium 24 h later and culture for 1–3 d.
5. Add TNF (100 U/mL) or IL-1 (5×10^{-11} g/mL) to cultures for desired period before migration assay (typically 4 h for neutrophils or 24 h for PBL).

3.2.5. Culturing HUVEC in Microslides (see **Note 7**)

3.2.5.1. PRETREATMENT WITH APES

1. Immerse microslides in nitric acid (50% v/v in distilled water) for 24 h (e.g., in batch ~100–300).
2. Wash thoroughly in beaker using running tap water and rinse through with deionised distilled water.
3. Blot water on tissue paper, and dry microslides at 37°C.
4. Place in polystyrene tubes and rinse twice with anhydrous acetone.

5. Immerse in a freshly prepared solution of APES (4% v/v in anhydrous acetone) for 1 min, ensuring all capillaries are filled (*see* **Note 8**).
6. Remove microslides from the APES and blot out onto tissue, ensuring that all capillaries are emptied.
7. Re-insert into a fresh aliquot of the solution for a further 1 min.
8. Remove APES by blotting and rinse the microslides once with anhydrous acetone, followed by three washes with deionized distilled water, and then dry at 37°C. Between each change, care must be taken to remove all the liquid from the microslides.
9. Autoclave the microslides at 121°C for 11 min and store aseptically, indefinitely.
10. All subsequent handling until the time of migration assay should be under sterile conditions.

3.2.5.2. Formation and Culture of Monolayer

1. Attach a short length of 2-mm ID silicon rubber tubing to APES-coated microslide. Draw in 1% gelatin (in PBS) using a pipettor with tip inserted into the tubing. Allow to coat for 30 min. Microslides hold approx 50 µL. The tubing also assists in handling and avoidance of fingerprints on microslides.
2. Trypsinize a single flask of HUVEC as in **Subheading 3.2.2.**
3. Make up cell pellet to only approx 500 µL with culture medium and transfer to one corner of a slightly tilted 35-mm Petri dish.
4. Aspirate approx 50 µL of cell suspension into each of six microslides using a pipettor inserted in the silicon tubing adaptor.
5. Place a sterile, glass microscope slide inside a 100-mm Petri dish, rest the filled microslides across it (to keep them horizontal), and incubate in the dish at 37°C for 1 h.
6. Prepare the special culture dish (**Subheading 2.3., item 5**) by attaching a length of silicon rubber tubing (40 cm long) onto each external arm. Add 50 mL of culture medium to the culture dish, prime the tubing, and clamp ends.
7. After 1 h settling/attachment, connect the microslides aseptically to the internal side-arms of the special culture dish, using sterile forceps.
8. Placed the dish in a humidified CO_2 incubator and pass the silicon tubing through a service port located in the incubator wall (manufactured to order; e.g., either model GA2000, LEEC Ltd., Nottingham, UK, or Nuaire DH, Triple Red, Thame, UK).
9. Attach the tubing to a multichannel, eight-roller pump (Watson Marlow 500 series pump with 308MC pumpheads; Watson-Marlow Bredel Pumps Ltd., Falmouth, UK), itself linked to a timed power supply (RS Components Ltd., Corby, UK).
10. Pump medium through microslides to waste (*see* **Fig. 1** for layout) at a flow rate of approx 0.2 mL/min for 30 s in each hour to change medium contained in the microslides.
11. The original seeding is designed to yield confluent monolayers within 24 h.
12. After 24 h, treat HUVEC with TNF (100 U/mL) or IL-1 (5×10^{-11} g/mL) for 4 h. Detach the microslides with tubing adaptor and place them in a separate disposable plastic culture dish. Aspirate cytokines (diluted in culture medium) into the separate microslides as desired, and repeat aspiration at hourly intervals.

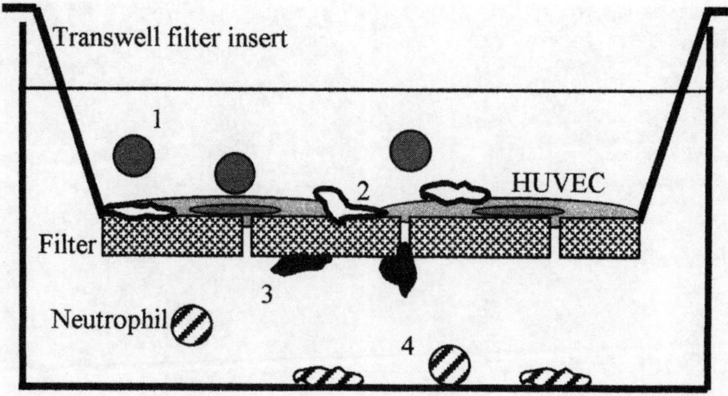

Fig. 3. Schematic representation of the Transwell assay. Neutrophils (2×10^6 cells/mL) are added into the upper chamber and allowed to interact with the TNF-stimulated endothelial cells (HUVEC). The neutrophils either remain nonadherent (1 = gray) or become attached to the surface of the HUVEC or migrate through them (2 = white). The neutrophils may migrate through the filter and either remain adherent to the basal surface of the filter (3 = black) or fully migrate into the lower chamber of the tissue culture plate (4 = hatched). Counting of cells retrieved from the upper and lower chambers determines the percentage of neutrophils that are nonadherent (1) or that fully transmigrate (4). Staining protocols allow counting and definition of the location of neutrophils within the filter/HUVEC layers (2 and 3).

3.3. Migration of Neutrophils
Through Endothelial Cells Cultured on Porous Filters

1. Stimulate HUVEC for 4 h with 0–100 U/mL TNF.
2. Remove TNF-containing medium from the upper and lower chamber.
3. Add 700 µL of fresh M199 + BSA to the lower chamber and 200 µL of neutrophils (2×10^6 cells/mL in M199 + BSA) to the upper chamber (*see* **Note 9**).
4. Allow the neutrophils to settle, adhere and migrate (*see* **Fig. 3**) at 37°C for desired time (e.g., 0.5, 2, or 24 h) (*see* **Note 10**).
5. Stop the experiment by transferring the filter into a fresh well.
6. Transfer the neutrophils from the upper chamber (above filter) into a fresh well.
7. Wash the upper chamber twice with 200 µL of M199 + BSA, and add washouts to the upper chamber samples. These represent the nonadherent neutrophils.
8. Wash the lower chamber once with 300 µL of M199 + BSA. Examine microscopically to ensure that all cells are removed and wash further if necessary. This represents the transmigrated neutrophils (*see* **Note 11**).
9. Fix the filter in 2% isotonic glutaraldehyde and store at 4°C.
10. Count the nonadherent and transmigrated samples using a Coulter counter to determine the number of nonadherent and transmigrated cells (*see* **Note 12**).

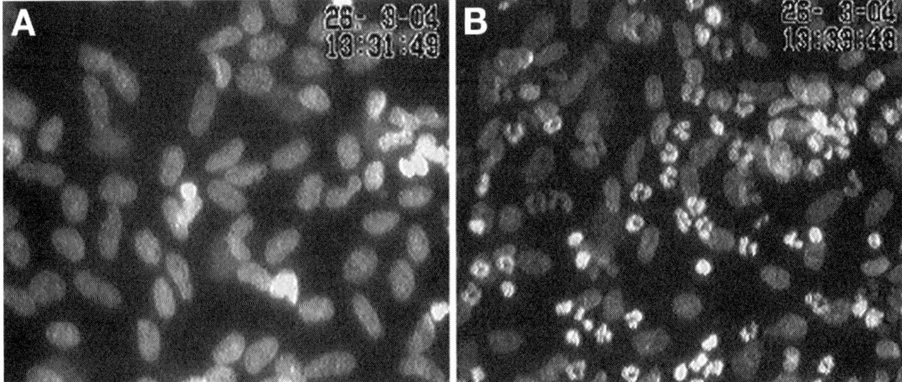

Fig. 4. Fluoresence micrographs of neutrophils adherent to HUVEC grown on Transwell filters. Neutrophils were settled for 120 min on HUVEC that were: **(A)** unstimulated; **(B)** treated with 100 U/mL tumor necrosis factor for 4 h. The filters were washed, fixed with 2% glutaraldehyde, and stained with 1 µg/mL bisbenzimide before being cut out and mounted on microscope slides. Oval nuclei represent HUVEC; smaller nuclei are neutrophils. The latter are clearly segmented when neutrophils spread on the surface.

11. From these data, the percentage of nonadherent, adherent (i.e., those remaining in the filter), and transmigrated neutrophils can be determined.
12. The location of the adherent cells on or in the filter can be assessed using fluoresence microscopy (*see* **Note 13** and **Fig. 4**).

3.4. Migration of Neutrophils Through Endothelial Cells Cultured in Multiwell Plates

3.4.1. Neutrophil Adhesion and Transmigration Assay

1. Stimulate HUVEC with 1, 10, or 100 U/mL TNF for 4 h.
2. Prewarm the microscope (*see* **Subheading 2.4.**) and PBS wash buffer to 37°C.
3. Rinse the HUVEC surface gently with 2 mL PBSA, using a plastic pipet, to remove any residual TNF.
4. Add 2 mL neutrophil suspension per well (time = 0 in **Fig. 5**).
5. Leave to settle for 5 min.
6. Aspirate off neutrophil suspension and gently rinse twice with PBSA (time = 5–7 min in **Fig. 5**) (*see* **Note 14**).
7. Add 2 mL PBSA and view HUVEC under phase contrast video-microscope with an objective magnification of ×20.
8. Follow the time scheme illustrated in **Fig. 5**: make video recordings immediately after the wash stage, at time = 7–9 min. Choose at least five different fields at random and record them for 15–20 s each. At time = 9 min, choose one field at random and record continuously for 15 min. At the end of this period (time = 24 min) choose another five fields to record until time = 26 min.

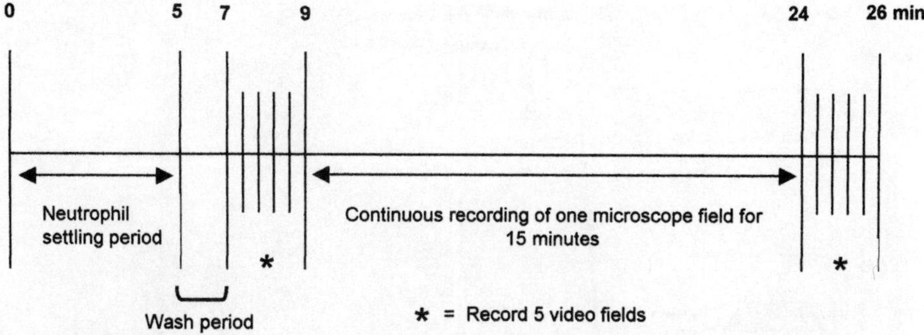

Fig. 5. Time line of the protocol for neutrophil adhesion to HUVEC in multiwell plate. Time elapsed from the initial addition of the neutrophil suspension to the endothelial monolayer is shown across the top.

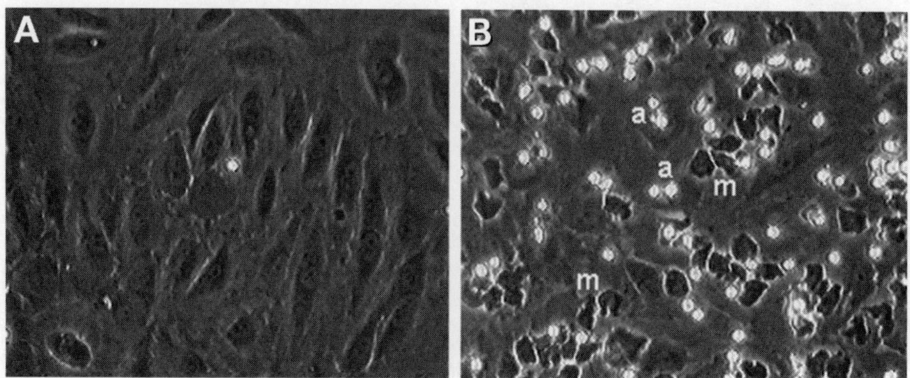

Fig. 6. Phase contrast micrographs of confluent monolayers of **(A)** untreated HUVEC or **(B)** HUVEC treated with 100 U/mL tumor necrosis factor after completion of a neutrophil adhesion assay. Neutrophils were settled for 5 min on HUVEC, and the nonadherent cells were washed off and allowed to migrate for a further 15 min. In **(B)**, phase-bright neutrophils are adherent to the surface of the HUVEC (a), and phase-dark spread cells (m) are migrated underneath the monolayer.

3.4.2. Data Analysis

1. Make video recordings of a microscope stage micrometer oriented parallel and perpendicular to flow. Use this to calibrate size of the video field observed on the monitor during playback. Also digitize images to calibrate the scale of the image analysis system.

2. To measure neutrophil adhesion: count all neutrophils visible in each video field (*see*, e.g., **Fig. 6**), e.g., recorded between 7–9 and 24–26 min (*see* **Note 15**). Take the average for the fields and convert to number per mm^2 using known dimensions of field. Multiply this by the area of the wells (9.6 cm^2 for six wells), and divide by the number of neutrophils added. Multiply by 100 to obtain the percentage of neutrophils adherent.

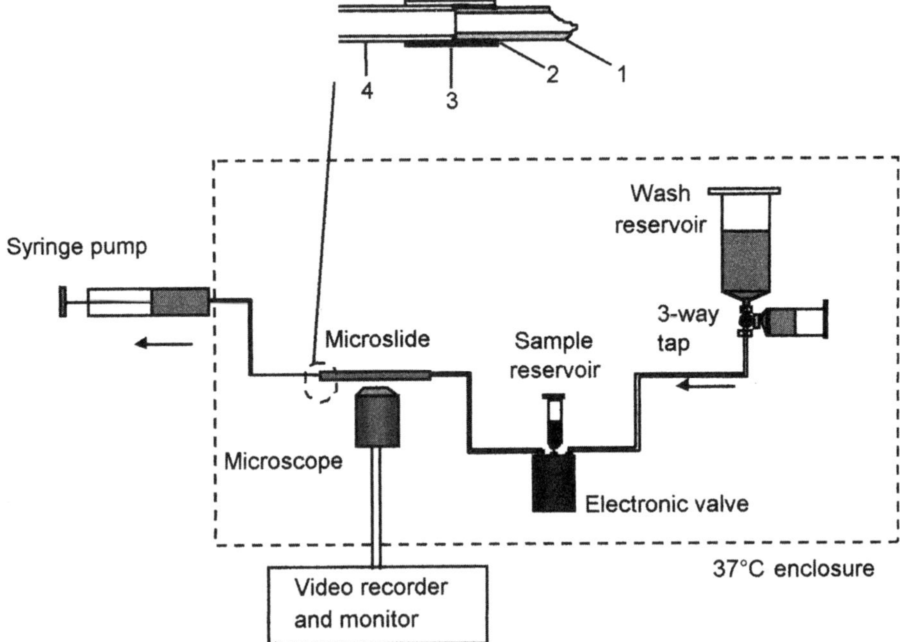

Fig. 7. Schematic of flow-based adhesion assay flow system. In expanded section, 1 = microslide, 2 = double-sided tape, 3 = silicon rubber tubing with ID/OD 2/4 mm, and 4 = silicon rubber tubing with ID/OD 1/2 mm.

3. To measure neutrophil transmigration: express the count of phase-dark neutrophils in each video field as a proportion of the total number of adherent cells (*see* **Note 15**). Take the average for the repeated fields. This can be done for both sets of recordings—min 7–9 and 24—to follow the progress of migration.

4. To measure the neutrophil migration velocity: digitize a sequence of images from the 15-min video sequence at 1-min intervals using a computer-assisted image analysis system, such as ImagePro. Outline individual neutrophils either on top of the HUVEC (phase-bright cells) or below the HUVEC (transmigrated cells) using a pointer to determine the positions of their centroids. Record the x and y coordinates of the centroids of the cells after each minute and calculate the distance moved each minute. Average this migration velocity over the desired period (*see* **Note 16**).

3.5. Flow-Based Assay of Neutrophil Migration

3.5.1. Setting Up the Flow Assay

1. Assemble flow system shown in **Fig. 7**, but without microslide attached. The electronic valve has a common output and two inputs, from "wash reservoir" and "sample reservoir," which can be selected by turning electronic valve on and off.

2. Fill wash reservoir with PBSA and rinse through all tubing, valves, and connectors with PBS, ensuring bubbles are displaced (e.g., using syringe attached to three-

way tap for positive ejection). Fill sample reservoir with PBSA and rinse through valve and attached tubing. Prime downstream syringe and tubing with PBS and load into syringe pump. All tubing must be liquid-filled.

3. Glue a coated microslide across the middle of a glass microscope slide using two spots of cyanoacrylate adhesive (Superglue Locktite UK, Welwyn Garden City, UK) applied to the edges of the slide. Discard tubing adaptor used for filling.
4. Wrap double-sided adhesive tape around each end of the microslide without obstructing lumen.
5. Connect microslide to silicon rubber tubing by pushing over each of the taped ends. Start at the upstream (sample) end to avoid injection of air. Squeeze the 2-mm ID silicon rubber tubing to flatten and ease over rectangular end of microslide, one corner first.
6. Place microslide onto microscope stage and start flow by turning on syringe pump in withdrawal mode, with electronic valve and three-way tap in position to allow delivery of PBS from wash reservoir.
7. Wash out culture medium (with cytokines) and observe surface using phase-contrast microscopy.
8. Adjust flow rate to that required for assay (*see* **Note 17**).
9. Perfusion of leukocytes is typically carried out at a flow rate ($Q = 0.395$ mL/min) equivalent to a wall shear rate of 140 per s and wall shear stress of 0.1 Pa (=1 dyn/cm^2).

3.5.2. Perfusing Leukocytes and Recording Behavior

1. Load isolated leukocytes into sample reservoir (**Fig. 7**) and allow to warm for 5 min.
2. Switch electronic valve so that leukocyte suspension is drawn through microslide.
3. Deliver timed bolus (e.g., 4 min). Typically, flowing leukocytes will be visible after about 30 s, the time required to displace dead volume in valve and tubing.
4. Switch electronic valve so that PBS from wash reservoir is perfused. Again, 30–60 s will be required before all leukocytes have been washed through the microslide.
5. Video recordings can be made as desired during inflow and washout of leukocytes. Typically, a series of fields should be recorded along the center line of the microslide during inflow (e.g., six fields recorded for 10 s each during the last minute of the bolus) for off-line analysis of the behavior (e.g., rolling or stationary adhesion) of the leukocytes. Another series should be made after 1-min washout (when the bolus is complete) for analysis of the numbers of adherent cells. Fields are recorded at later times (e.g., after a further 5 and 10 min) to assess the progress of migration through the monolayer. In addition, a single field may be recorded between times so that the migration of individual cells can be tracked and their velocities analyzed. Analyses are made off-line. If a defined timing protocol is developed, digital images or sequences of digital images can be recorded instead of video images. The continual recording of the latter gives flexibility in analysis.
6. To study inhibitory antibodies pretreat either leukocytes or endothelial monolayers for 20 min before perfusion. If competitive inhibitors are to be used (e.g., RGD peptide), they need to be present in media throughout, otherwise they will be washed out.

3.5.3. Analysis of Leukocyte Behavior From Video Recordings

1. Calibrate size of video fields and image analysis system as in **Subheading 3.4.2.**
2. Digitize a sequence of images at 1-s intervals from recordings made at desired times.
3. Count cells present on a stop-frame video field at the start of a sequence. Repeat and average counts for the series of sequences recorded, e.g., after washout of non-adherent cells. Convert this to count of adherent cells/mm^2.
4. Divide this by the numbers of leukocytes perfused (in units of 10^6 cells) to obtain number adherent/mm^2/10^6 perfused. The number perfused is simply calculated by multiplying the concentration of the suspension (usually 10^6/mL) by the flow rate (e.g., 0.395 mL/min) by the duration of the bolus (e.g., 4 min). This normalization allows correction for changes in conditions (bolus duration, cell concentration, flow rate) between experiments and effectively calculates an efficiency of adhesion.
5. The count includes cells that are rolling (circular phase bright cells tumbling slowly at ~1–10 µm/s over the surface) or stably adherent on the endothelial surface (phase-bright cells typically with distorted outline and migrating slowly on the surface at <10 µm/min) or transmigrated cells (phase-dark spread cells migrating under the HUVEC at >10 µm/min). Nonadherent cells will only be visible as blurred streaks.
6. To assist in obtaining data for proportion of adherent cells behaving in the different ways, play the digitized sequence as a loop. Observing cells in turn, it is easy to classify them as rolling, stabley adherent, or transmigrated.
7. Repeat the analysis at different times (e.g., after 1, 5, or 10 min of washout) to quantify the progress of migration through the endothelium or any changes in behavior such as transmigration through the endothelial monolayer.
8. To measure rolling velocity, mark the leading edges of a series of cells to be followed and move to second captured frame. Remark the leading edges and record the distance moved. Repeat through the 10-s sequence. This will yield data for position vs time. Velocity for each cell can be averaged over the observation time and estimates of variation in velocity made if desired.
9. To measure migration velocity in extended video sequences, images are digitized at 1-min intervals over 5–10 min. The rate of migration is measured from the digitized images as in **Subheading 3.4.2.**
10. In the presence of flow, cells on the endothelial surface may migrate preferentially along the flow axis *(11)*. The direction of migration can be quanitifed as the net distance moved in a chosen direction (e.g., parallel with the flow axis) divided by the sum of the individual distances moved in the separate minutes. A value of 1 represents movement in a straight line along the chosen axis. A value of 0 means no net displacement in that direction.

4. Notes

1. There are various methods for isolating leukocytes from blood, and **Subheading 3.1.** describes a simple one that we use regularly. The cells should appear spherical. If the neutrophils have distorted shape, one must suspect unintentional activa-

tion. It is advisable also to test viability of preparations (e.g., ~99% viable judged with Trypan blue) and purity (e.g., 95% polymorphonuclear when isolating neutrophils judged with crystal violet), especially when first starting this type of work. Lymphocytes prepared in this way will still have some monocyte contamination. Further purification of lymphocyte subsets can be made using immunomagnetics selection (e.g., Dynabeads, Dynal Biotech UK, Bromborough, UK; MACS, Miltenyi Biotec Ltd., Bisley, UK). Purified monocytes can be obtained using a different type of density gradient, e.g., as described in **ref. *19***.

2. PBSA is sufficient to maintain a viable HUVEC monolayer over a short period of time. However, PBSA is unable to maintain an intact HUVEC monolayer following 24 h of culture as judged by visual observations and a decrease in electrical resistance across the monolayer. Lymphocytes can be suspended in endothelial culture medium, but in our experience neutrophils become activated in the presence of fetal calf serum. Comparing RPMI-1640, M199, and hybridoma solution (Sigma) supplemented with 0.15% BSA, the M199 + BSA maintained endothelial morphology and electrical resistance and had the least effect on neutrophil morphology.

3. Previous studies have precoated the Transwell filters with collagen or fibronectin (FN) *(3–5,20)*. It has been suggested that this coating increases the percentage of leukocyte migration. However, comparing uncoated filters with FN-coated filters (either coated before the assay with 20 µg/mL human plasma FN (Sigma) or bought precoated with 170–200 µg/mL FN from BD), we found no significant differences in the percentages of neutrophils transmigrating.

4. The Transwell assay can also be performed using six-well format filters. In this case, add 3 mL of medium to the lower chamber and 2 mL of HUVEC-containing solution to the upper chamber. Comparable data are achieved using either the 24- or 6-well format, but an advantage of the 6-well format is that the increased numbers of leukocytes added and retrieved allows one to study the phenotypes of the transmigrated leukocytes (e.g., preferential migration of lymphocyte subsets) using immunofluorescence and flow cytometry.

5. HUVEC form a confluent monolayer within 24 h at this seeding density. However, electrical resistance across the monolayer continues to increase for about 4 d. We compared the ability of neutrophils to transmigrate through day 1 or day 4 HUVEC with or without stimulation with TNF. A significantly lower percentage of neutrophils transmigrated through unstimulated day 4 HUVEC compared with day 1 HUVEC.

6. One confluent 25-cm^2 flask of HUVEC, resuspended in 6 mL, will seed three wells (2 mL per well) to produce a confluent monolayer within 24 h. Alternatively, one 25-cm^2 flask can be resuspended in 8 mL and used to seed four wells, which will be confluent in 2–3 d. We have found no difference in adhesion or migration on 1- or 3-d cultures

7. The microslide culture system is one we have developed for both culture under flow and testing of adhesion and migration. A system is available commercially (GlycoTech, Rockville, MD) based on culture of endothelial cells in 35-mm-diameter culture dishes. An adaptor and gasket are inserted on top of the cells, which

allows perfusion of fluid, (± suspended leukocytes) over the endothelial surface. We have no experience with this system, but its use for adhesion assays has been described *(12)*.

8. Successful coating with APES requires that the reagent be anhydrous. We buy small volumes adequate to coat a batch of microslides, use a fresh bottle for each batch, and discard any unused reagent. It is important to efficiently remove all liquid from each microslide between changes and to ensure all bubbles are displaced on refilling with agents.

9. The assay can be used to assess the migration of lymphocytes, either PBL or isolated subtypes. HUVEC are stimulated with TNF for 24 h prior to the start of the assay. Lymphocytes (2×10^6 cells/mL in M199 + BSA) are allowed to migrate for 20 h. Lymphocytes are slower at migrating through filters than neutrophils *(21–23)*.

10. While optimizing this protocol, we compared neutrophil migration after 0.5, 2, and 24 h. Transmigration increased significantly with time, and we routinely use a 2-h period.

11. If adhesion of neutrophils to the bottom of the well causes problems, it can be precoated with the nonadhesive substrate 50 mg/mL polyHEMA *(24,25)*.

12. There are alternative methods of analyzing neutrophil counts including prelabeling the neutrophils with fluorescent dyes or radiolabeled isotopes or measuring myeloperoxidase activity. Also, BD Biosciences sells a Transwell filter that has a patented light-tight PET membrane that efficiently blocks the transmission of light within the range of 490–700 nm (www.bdbiosceinces.com). Using this system, the number of transmigrated fluorescent cells beneath the filter can be analyzed using a fluorescence plate reader.

13. To assess the location of adherent cells:
    a. Stain the glutaraldehyde fixed filter with the nuclear dye 1 µg/mL bisbenzimide (Sigma) for 15 min in the dark.
    b. Wash the filter four times in PBS (this is important because otherwise, the pores hold the fluorescent dye).
    c. Cut the filter out, using a scalpel, directly onto a microscope slide.
    d. Mount with antifade agent (e.g., DABCO; Sigma).
    e. Using a fluoresence microscope with UV illumination and ×40 objective, neutrophils adherent to the upper surface can be counted (*see*, e.g., **Fig. 4**).
    The nuclei of the HUVEC will also be visible. Next, move the focus down through the filter (a distance of ~10 µm) until transmigrated neutrophils adhered to the back of the filter come into view and count these cells. The proportion of adherent neutrophils either side of the filter can be calculated.

14. This short settling time and effective wash procedure in six-well plates results in little or no background neutrophil adhesion on unstimulated controls (**Fig. 6A**).

15. The adherent neutrophils can be classified into two groups: (1) phase-bright cells adherent to the upper surface of the endothelial cells and (2) phase-dark spread cells that are transmigrated under the endothelial monolayer (*see* **Fig. 6B**). Neutrophil adhesion levels should not change between the two sets of recordings—only the proportion that have transmigrated.

16. Our studies show no significant variation in migration velocity with time, but interestingly, transmigrated cells may migrate more rapidly under the HUVEC than those on the endothelial surface.

17. The flow rate (Q) required to give a desired wall shear rate (γ_w in s^{-1}) or wall shear stress (τ_w in Pascal, Pa) is calculated from the internal width (w) and internal depth (h) of the microslide and the viscosity (n) of the flowing medium using the formulas:

$$\gamma = (6.Q)/(w \cdot h^2)$$

$$\tau = n \cdot \gamma$$

Because $w = 3$ mm, $h = 0.3$ mm, $n = 0.7$ mPa.s for PBS at 37°C, this can be manipulated to give:

$$Q \text{ (mL/min)} = 0.0027 \cdot \gamma_w \text{ (s}^{-1})$$

or

$$Q \text{ (mL/min)} = 3.95 \cdot \tau_w \text{ (Pa)}$$

References

1. Springer, T. A. (1995) Traffic signals on endothelium for lymphocyte recirculation and leukocyte emigration. *Ann. Rev. Physiol.* **57,** 827–872.
2. Johnson-Leger, C., Aurrand-Lions, M., and Imhof, B. A. (2000) The parting of the endothelium: miracle or simply a junctional affair. *J. Cell Sci.* **113,** 921–933.
3. Lampugnani, M. G., Resnati, M., Raiteri, M., et al. (1992) A novel endothelial-specific membrane protein is a marker of cell-cell contacts. *J. Cell Biol.* **118,** 1511–1522.
4. Kuijpers, T. W., Hakkert, B. C., Hart, M. H. L., and Roos, D. (1992) Neutrophil migration across monolayers of cytokine-prestimulated endothelial cells: a role for platelet-activating factor and IL-8. *J. Cell Biol.* **117,** 565–572.
5. Cooper, D., Lindberg, F. P., Gamble, J. R., Brown, E. J., and Vadas, M. A. (1995) Transendothelial migration of neutrophils involves integrin-associated protein (CD47). *Proc. Natl. Acad. Sci. USA* **92,** 3978–3982.
6. Everitt, E. A., Malik, A. B., and Hendey, B. (1996) Fibronectin enhances the migration rate of human neutrophils in vitro. *J. Leuk. Biol.* **60,** 199–206.
7. Burns, A. R., Walker, D. C., Brown, E. S., et al. (1997) Neutrophil transendothelial migration is independent of tight junctions and occurs preferentially at tricellular corners. *J Immunol.* **159,** 2893–2903.
8. Luu, N. T., Rainger, G. E., and Nash, G. B. (1999) Kinetics of the different steps during neutrophil migration through cultured endothelial monolayers treated with tumour necrosis factor-alpha. *J. Vasc. Res.* **36,** 477–485.
9. Rainger, G. E., Fisher, A., Shearman, C., and Nash, G. B. (1995) Adhesion of flowing neutrophils to cultured endothelial cells after hypoxia and reoxygenation in vitro. *Am. J. Physiol.* **269,** 1398–1406.
10. Luu, N. T., Rainger, G. E., Buckley, C. D., and Nash, G. B. (2003) CD31 regulates direction and rate of neutrophil migration over and under endothelial cells. *J. Vasc. Res.* **40,** 467–479.

11. Rainger, G. E., Buckley, C. D., Simmons, D. L., and Nash, G. B. (1999) Neutrophils sense flow-generated stress and direct their migration through alphaVbeta3-integrin. *Am. J. Physiol.* **276,** 858–864.

12. Cinamon, G., Shinder, V., and Alon, R. (2001) Shear forces promote lymphocyte migration across vascular endothelium bearing apical chemokines. *Nat. Immunol.* **2,** 515–522.

13. Oppenheimer-Marks, N. and Lipsky, P. E. (1997) Migration of naive and memory T cells. *Immunol. Today* **18,** 456–457.

14. Oppenheimer-Marks, N. and Ziff, M. (1988) Migration of lymphocytes through endothelial cell monolayers: augmentation by interferon-gamma. *Cell. Immunol.* **114,** 307–323.

15. Borthwick, N. J., Akbar, A. N., MacCormac, L. P., et al. (1997) Selective migration of highly differentiated primed T cells, defined by low expression of CD45RB, across human umbilical vein endothelial cells: effects of viral infection on transmigration. *Immunology* **90,** 272–280.

16. Allport, J. R., Muller, W. A., and Luscinskas, F. W. (2000) Monocytes induce reversible focal changes in vascular endothelial cadherin complex during transendothelial migration under flow. *J. Cell Biol.* **148,** 203–216.

17. Ades, E. W., Candal, F. J., Swerlick, R. A., George, V. G., Summers, S., Bosse, D. C., and Lawley, T. J. (1992) HMEC-1: establishment of an immortalized human microvascular endothelial cell line. *J. Invest. Dermatol.* **99,** 683–690.

18. Cooke, B. M., Usami, S., Perry, I., and Nash, G. B. (1993) A simplified method for culture of endothelial cells and analysis of adhesion of blood cells under conditions of flow. *Microvasc. Res.* **45,** 33–45.

19. Tsouknos, A., Nash, G. B., and Rainger, G. E. (2003) Monocytes initiate a cycle of leukocyte recruitment when cocultured with endothelial cells. *Atherosclerosis* **170,** 49–58.

20. Everitt, E. A., Malik, A. B., and Hendey, B. (1996) Fibronectin enhances the migration rate of human neutrophils in vitro. *J. Leukocyte Biol.* **60,** 199–206.

21. Oppenheimer-Marks, N. and Lipsky, P. E. (1997) Migration of naive and memory T cells. *Immunol. Today* **18,** 456–457.

22. Oppenheimer-Marks, N. and Ziff, M. (1988) Migration of lymphocytes through endothelial cell monolayers: augmentation by interferon-gamma. *Cell. Immunol.* **114,** 307–323.

23. Borthwick, N. J., Akbar, A. N., MacCormac, L. P., et al. (1997) Selective migration of highly differentiated primed T cells, defined by low expression of CD45RB, across human umbilical vein endothelial cells: effects of viral infection on transmigration. *Immunology* **90,** 272–280.

24. Kettritz, R., Xu, Y. X., Kerren, T., Quass, P., Klein, J. B., Luft, F. C., and Haller, H. (1999) Extracellular matrix regulates apoptosis in human neutrophils. *Kidney Int.* **55,** 562–571.

25. Folkman, J. and Moscona, A. (1978) Role of cell shape in growth control. *Nature* **273,** 345–349.

5

Biochemical Purification
of Pseudopodia from Migratory Cells

Yingchun Wang and Richard L. Klemke

Summary

Cell migration requires the formation of a leading pseudopodium (lamellipodium) in the direction of movement. This process requires signal amplification to facilitate directional sensing mechanisms that lead to actin-mediated membrane extension. However, it has been difficult to study pseudopodia formation because it has not been possible to purify this structure for biochemical analysis. Here we describe a method to biochemically purify the protruding pseudopodium from the cell body compartment using polycarbonate microporous filters. Cells are cultured on top of 3.0-μm porous filters and allowed to extend pseudopodia through the small pores to the undersurface in response to a gradient of either chemokine or extracellular matrix (ECM) protein. Pseudopodia and cell bodies are then differentially scraped from the filter surface into lysis buffer for biochemical analysis. Using this method, it is possible to identify novel pseudopodium and cell body proteins as well as study the spatiotemporal organization of signaling processes that regulate pseudopodium formation and cell polarity. This method will help facilitate our understanding of how cells protrude pseudopodia through small openings in the ECM and vasculature during cancer cell invasion, immune cell surveillance, and embryonic development.

Key Words: Chemokine; cell migration; chemotaxis; haptotaxis; pseudopodium; cytoskeleton; signal transduction; cell metastasis; focal adhesions; lamellipodium.

1. Introduction

Cell migration plays a central role in embryonic development, wound healing, immune function, and tumor invasion and metastasis *(1,2)*. In many cases cells initiate migration by sensing a gradient of either chemokine or extracellular matrix (ECM) proteins, which serve as an adhesive substrate *(2–4)*. Setting the migratory compass is achieved by gradient sensing mechanisms that give

From: *Methods in Molecular Biology, vol. 370:*
Adhesion Protein Protocols, Second Edition
Edited by: A. S. Coutts © Humana Press Inc., Totowa, NJ

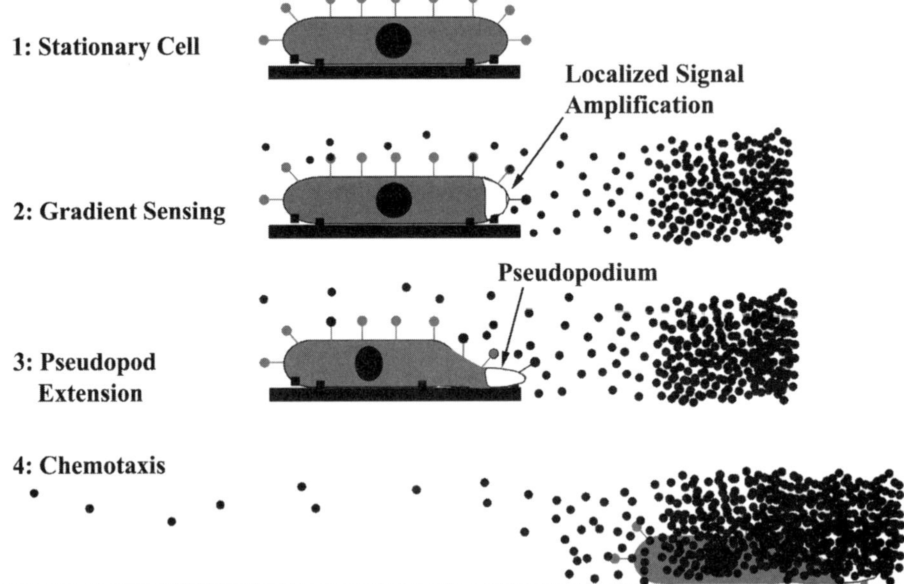

Fig. 1. Model illustrating the basic steps of cell polarization and chemotaxis. Step 1: stationary cells are attached to the underlying extracellular matrix (ECM) through integrin receptors (squares). Step 2: cells are then exposed to a soluble gradient of chemokine or growth factor. This activates cell surface chemoattractant receptors leading to localized activation and amplification of signals on the side facing the gradient. Step 3: the amplified signals then regulate localized actin dynamics and lead to membrane protrusion in the direction of the gradient. This process is independent of actual cell body translocation or chemotaxis and marks the first sign of morphological polarity with establishment of a dominant leading pseudopodium and posterior compartment. Step 4: once a dominant pseudopodium is established, the cell begins to migrate in the direction of the gradient as it undergoes repeated cycles of membrane protrusion, adhesion to the ECM, and release of old attachment in the rear.

rise to localized amplification of signals on the side facing the gradient *(3,5–8)* (**Fig. 1**). This response drives actin polymerization and protrusion of a leading pseudopodium (lamellipodium) in the direction of the stimulus *(9–11)*. Cell translocation is then achieved by the coordinated protrusion of a pseudopodium followed by retraction of the rear compartment of the cell (cell body) (**Fig. 1**). Although it is clear that cells need to establish morphological polarity to translocate, studying this process at the molecular level has been hindered because it has not been possible to isolate the pseudopodium for biochemical analysis *(9)*.

Recently we introduced a novel approach to biochemically purify the protruding pseudopodium from the cell body using microporous filters *(3)*. We have

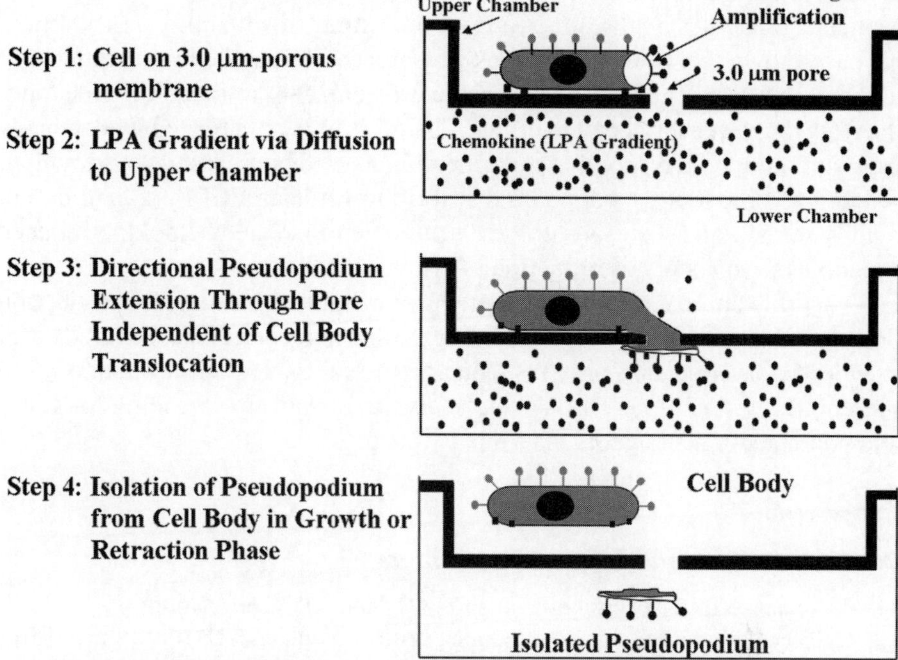

Fig. 2. Schematic showing biochemical isolation of pseudopodia. Step 1: migratory cells are allowed to attach and spread on the upper surfaces of 3.0-µm porous polycarbonate filters coated with extracellular matrix proteins on both upper and lower surfaces in a Costar chamber. Step 2: cells are then stimulated with chemoattractant, which is placed in the lower chamber and establishes a gradient by diffusing through the 3.0-µm pores. Step 3: Cells begin to extend pseudopodia in the direction of the chemoattractant gradient through the 3.0-µm pores, and the posterior cell body compartments containing nuclei remain on the upper surface of the filter. Step 4: The pseudopodia can be isolated from cell body in growth or retraction phase as described in **Heading 3.**

shown that cells extend pseudopodia through small (3.0-µm) pores of a polycarbonate filter in response to either a chemokine or ECM gradient *(3,4)* (**Fig. 2**). Confocal microscopy and time-lapse imaging of this process demonstrates that cells extend pseudopodia only through the pore toward the gradient cue and not randomly on top of the filter *(3)*. Although each cell achieves polarity by extending a membrane protrusion through the small opening in the filter, the cell body does not translocate through the pore. The cell bodies on the upper surface can then be sheared off with a cell scraper, leaving the isolated pseudopodia on the lower surface. Pseudopodia are then harvested into the appropriate buffer by scraping the lower surface of the filter with a cell scraper. The cell bodies are harvested in a similar manner: pseudopodia on the lower surface are removed first and cell bodies on the upper surface are extracted with buffer. In addition, we have shown that when the chemokine is removed from the lower chamber or added to the upper chamber to disrupt the gradient, cells retract their pseudopodia from the lower surface of the filter. Therefore, it is possible to biochemically purify pseudopodia responding to a chemokine gradient in a state of growth or retraction. Note that pseudopodia that form towards an ECM gradient do not readily retract, and therefore it is only possible to isolate chemokine-induced pseudopodia in a state of retraction *(4)*.

Using this approach, we and others have examined the signal transduction processes that control pseudopodia formation and retraction and uncovered novel pseudopodial-associated proteins using large-scale, mass spectrometry-based proteomics *(3,4,12–14)*. This system is also amendable to screening for genes and pharmacological agents that impact this process.

2. Materials

2.1. Cell Culture and Microporous Filters

1. Cell lines: NIH 3T3 fibroblasts and COS-7 from ATCC (*see* **Note 1**).
2. Dulbecco's modified Eagle's medium (DMEM) complete: DMEM (Gibco/BRL, Bethesda, MD) supplemented with 10% fetal bovine serum (FBS; Gemini Bio-Products, Woodland, CA), 2 mM L-glutamine (Sigma, St. Louis, MO), 50 µg/mL gentamicin (Sigma), and 1 mM sodium pyruvate (Sigma).
3. Starvation medium: same as DMEM complete but without FBS (*see* **Note 2**).
4. Migration/Adhesion medium: same as DMEM complete but without FBS and with 0.25% (W/V) radioimmunoassay-grade Fraction V bovine serum albumin (Sigma).
5. L-α-Lysophosphatidic acid, Oleoyl, sodium (LPA; Sigma) is dissolved in water at a concentration of 1 mg/mL and stored in aliquots of 10 µL at −80°C for up to 1 yr.
6. Human fibronectin (BD Biosciences) is dissolved in sterile MilliQ water at 1 mg/mL and can be kept at 4°C for 2 mo (*see* **Note 3**).
7. 1X Trypsin–ethylene diamine tetraacetic acid solution (Gibco).
8. Sterile phosphate-buffered saline (PBS) (1X).
9. Parafilm (Fisher).
10. Cotton swab (or cotton-tipped cleaning sticks) (Puritan Medical Products Company LLC, Guilford, ME).
11. Cell scrapers (Corning).
12. 1% Sodium dodecyl sulfate (SDS) lysis buffer: 1% (w/v) SDS, 2 mM sodium ortho-vanadate, 1 mM phenylmethylsulfonyl fluoride (Sigma), and half of a tablet of protease inhibitors (Roche) per 10 mL. Aliquot to 1 mL and store at −80°C.
13. 100% Ice-cold methanol.
14. Heat block at 100°C.
15. 24-mm Transwell containing polycarbonate membrane with 3.0-µm pores in six-well plate format (Corning) (*see* **Note 4**).

16. Micro BCA protein assay kit (Pierce Biotechnology, Rockford, IL).

2.2. Pseudopodia Staining and Imaging

1. Crystal violet powder (Sigma).
2. Boric acid 0.2 M stock (Solution A) (Fluka).
3. Disodium tetraborate 0.05 mM stock (Solution B) (Sigma).
4. Ethanol (100%).
5. 4% Paraformaldehyde (Electron Microscopy Sciences, Ft. Washington, PA).
6. Rodamine-conjugated phalloidin (Sigma).
7. Fluoromount (Southern Biotechnology Associates, Inc., Birmingham, AL).

3. Methods

The methods outlined below describe how to purify pseudopodia responding to either a soluble chemokine gradient (chemotaxis) or immobilized ECM gradient (haptotaxis) coated on the undersurface of microporous filters. The proper extension of pseudopodia through small micropores depends on three critical factors. First, cells must readily attach to the ECM protein-coated filter. Different cell lines will require different ECM proteins for proper attachment, and this needs to be determined for each new cell line tested. Second, the diffusion gradient must be established correctly. Shaking or rough handling of the chambers will disrupt the diffusion gradient. Once the chemokine is placed in the lower chamber, it is imperative that the plate be quickly but gently placed into the incubator and not disturbed until it is time to harvest the pseudopodia. It is also important to ensure that the microporous filters do not leak because this will create a nonuniform gradient and result in little or no pseudopodia extension. Finally, the optimal concentration of chemokine or ECM must be determined for each new chemokine/ECM protein tested.

Once cells have been properly stimulated, cell bodies and/or pseudopodia can be harvested and quantified either by using a standard protein determination assay or by staining with crystal violet dye and reading on an enzyme-linked immunosorbent assay (ELISA) plate reader. Alternatively, the filters can be cut from the plastic chamber, stained with antibodies and compounds, and mounted on glass slides to allow visualization of the cell compartments in high resolution using either standard or confocal microscopy.

3.1. Preparation of Cells for Chemokine and ECM Gradient Pseudopodia Assays

1. For NIH 3T3 and COS-7 cells, passage three 100-mm Petri dishes of cells (60–70% confluent) into 12 100-mm dishes containing 7–9 mL of DMEM complete medium. This will yield enough cells for approx 12 24-mm chambers and 400–500 µg of purified pseudopodia protein.
2. Culture cells for 3 d until cells are 70–80% confluent.

3. The day before the pseudopodia assay, carefully aspirate off the DMEM complete medium and add 7 mL of starvation medium to the cells. Starve cells overnight.

3.2. Coating of Microporous Filters for Chemokine and ECM Gradient Pseudopodia Assays

1. Prepare a 5 µg/mL solution of fibronectin in sterile PBS without cations (*see* **Note 5**). Check that the filters do not leak (*see* **Note 6**).
2. For chemokine gradient pseudopodia assays (chemotaxis), add 2 mL of fibronectin solution to the bottom chamber and 1.3 mL to the top chamber. Make sure that fluid evenly coats the filter surface.
3. For ECM gradient pseudopodia assays (haptotaxis), add 2 mL of fibronectin solution to the bottom chamber. Most cells will attach to the polycarbonate filter on the upper surface and extend pseudopodia toward the fibronectin coating on the bottom. No fibronectin is placed in control chambers (little or no pseudopodia formation should occur).
4. Allow fibronectin to coat the chamber(s) for 2 h at 37°C.

3.3. Attachment of Cells to Microporous Filters for Chemokine and ECM Gradient Pseudopodia Assays

1. Approximately 40 min before the chambers are finished coating with fibronectin (**step 4**, above), remove the cells from the dish by aspirating off the starvation medium, rinsing in PBS, and then adding 2 mL 1X trypsin-ethylene diamine tetraacetic acid solution to each dish. Incubate at 37°C for 3–4 min and then either gently tap or apply a stream of PBS with a pipet to dislodge the cells.
2. Add 3 mL migration/adhesion (37°C) medium per dish to quench the trypsin.
3. Collect the cells into two 50-mL tubes and spin for 3 min at 150g to pellet the cells. Carefully aspirate the supernatant without disturbing the cell pellet and then add 10 mL of migration/adhesion medium to each tube to resuspend the cells. Combine the cells into one 50-mL tube.
4. Count the cells and make the appropriate dilution with migration/adhesion medium to make the final concentration 1×10^6 cells/mL (*see* **Note 7**).
5. Add 2.5 mL of migration/adhesion medium to each well (lower chamber) of two six-well dishes.
6. Remove the microporous filters from the fibronectin solution and gently shake off the excess liquid (*see* **Note 8**).
7. Add 1.5 mL of cell suspension (1.5×10^6 cells) to the upper chamber and immediately place into a well of a six-well dish containing 2.5 mL of migration/adhesion medium as prepared above (**Subheading 2.1.**). Repeat for each of the filters.
8. For chemokine gradient pseudopodia assays (chemotaxis), allow the cells to attach and spread for 2–2.5 h at 37°C in a tissue culture incubator. Cell adhesion and spreading can be examined using a standard phase contrast microscope (*see* **Note 9**). Prepare growth or retraction phase pseudopodia as described below (*see* **Subheading 3.4.**).

9. For ECM gradient pseudopodia assays (haptotaxis), allow the cells to extend pseudopodia through the pores for 30–60 min at 37°C in a tissue culture incubator (*see* **Note 9**). Harvest pseudopodia and cell bodies as described below (*see* **Subheadings 3.5.** and **3.6.**, respectively). Alternatively, prepare filters for visualization/quantification (*see* **Subheading 3.8.**).

3.4. Stimulation of Pseudopodia Growth and Retraction for Chemokine Gradient Assays

1. Add 2.5 mL of migration/adhesion medium containing 100 ng/mL LPA to the lower chamber of each well of two new six-well dishes (*see* **Note 10**). No LPA is placed in control chambers (little or no pseudopodia formation should occur).
2. Remove the upper chamber with attached cells and immediately place it into a well containing the LPA solution. Be careful not to shake or bang the dish because this can disrupt the LPA diffusion gradient between the upper and lower chamber (*see* **Note 11**). It is also important to remove any bubbles that may have been introduced underneath the filter by gently tilting the chamber to one side.
3. Repeat **step 2** for each filter.
4. Gently place dishes into the incubator and allow pseudopodia to grow for approx 60 min (*see* **Note 12**).
5. For growth phase pseudopodia, harvest pseudopodia and cell bodies as described below (*see* **Subheadings 3.5.** and **3.6.**, respectively). Alternatively, prepare filters for visualization/quantification (*see* **Subheading 3.8.**).
6. For retraction phase pseudopodia, add 2.5 mL of migration/adhesion medium to each well of a new six-well dish and place it in the incubator. The number of chambers needed will depend on the amount of protein needed.
7. Remove the upper chamber after the appropriate chemokine stimulation (**step 4**) and gently aspirate off the medium from the upper and lower surfaces of the chamber being careful not to disrupt the cells.
8. Place filter directly into a well of a six-well dish containing 2.5 mL of migration/adhesion medium (**step 6**). There is no need to add buffer to the top of the filter because the buffer from the lower chamber will rapidly diffuse to the upper chamber.
9. Allow pseudopodia to retract for the appropriate time (typically 30 min for COS-7 and NIH 3T3 cells). Harvest pseudopodia and cell bodies as described below (*see* **Subheadings 3.5.** and **3.6.**, respectively). Alternatively, prepare filters for visualization/quantification (*see* **Subheading 3.8.**).

3.5. Biochemical Purification of Pseudopodia

1. Remove filter from dish and quickly but gently rinse in PBS at room temperature. Place the filter directly into ice-cold 100% methanol and fix for 30–60 min (*see* **Note 13**).
2. Flatten out a piece of parafilm on the top of a glass plate and place a 200-μL drop of 1% SDS lysis buffer on the parafilm (*see* **Note 14**).
3. Remove filter from fixative and gently rinse in PBS at room temperature.

4. Shake off excess PBS, and then use a cotton swab with the tip flattened to remove cell bodies from the top of the filter. It is important that all cells and debris be removed from the top, especially around the edges where the filter attaches to the plastic chamber.
5. Rinse in PBS to remove cell debris.
6. Repeat **step 4**.
7. Cut the filter along the outer edge of the chamber using a sharp razor blade. This should generate a round filter with a smooth edge (*see* **Note 15**).
8. Place the filter prepared in **step 7** onto the lysis buffer with the pseudopodia facing down and allow the proteins to solubilize for 30 s.
9. Using a cell scraper, scrape pseudopodia from the filter into the lysis buffer. Using a pipet, carefully retrieve the residue buffer and protein left on the filter and add it back to the drop of lysis buffer on the parafilm (*see* **Note 16**).
10. Repeat **steps 3–9** for each filter of a six-well dish. Use 200 µL of lysis buffer (step 2) for each six-well dish. The pseudopodia protein is now concentrated in a small volume of lysis buffer.
11. Transfer the protein lysate to a 1.5-mL centrifuge tube with a pipette. Typically, 300–350 µL of lysate containing 400–500 µg of protein is recovered from 12 filters.
12. Immediately boil the sample for 10 min (*see* **Note 17**).

3.6. Biochemical Purification of Cell Bodies

1. Remove filter from dish and quickly but gently rinse in PBS at room temperature. Place the filter directly into ice-cold 100% methanol and fix for 30–60 min (*see* **Note 13**).
2. Remove the chamber from the fixative and gently rinse in PBS at room temperature.
3. Shake off excess PBS and manually remove pseudopodia from the bottom of the filter using a cotton swab. It is important that all debris be removed from the bottom prior to protein extraction in lysis buffer.
4. Rinse in PBS to remove debris.
5. Repeat **step 3**.
6. Add 200 µL of lysis buffer to each well.
7. Using a pipet, retrieve the lysis buffer from each well and combine it in a 1.5-mL tube.

3.7. Determining Protein Concentration in Cell Bodies and Pseudopodia

1. Once the pseudopodia and cell body proteins are prepared as described above (*see* **Subheadings 3.1.** to **3.6.**), the protein concentration can be determined using the micro BCA kit according to the manufacturer's recommended methodology.
2. To determine the relative expression level of a protein(s) in each fraction, we load equivalent amounts of cell body and pseudopodia proteins (20–30 µg) for comparison using standard SDS–polyacrylamide gel electrophoresis and Western blotting methods (**Fig. 3**).
3. The lysates can be stored at −80°C for at least a year.

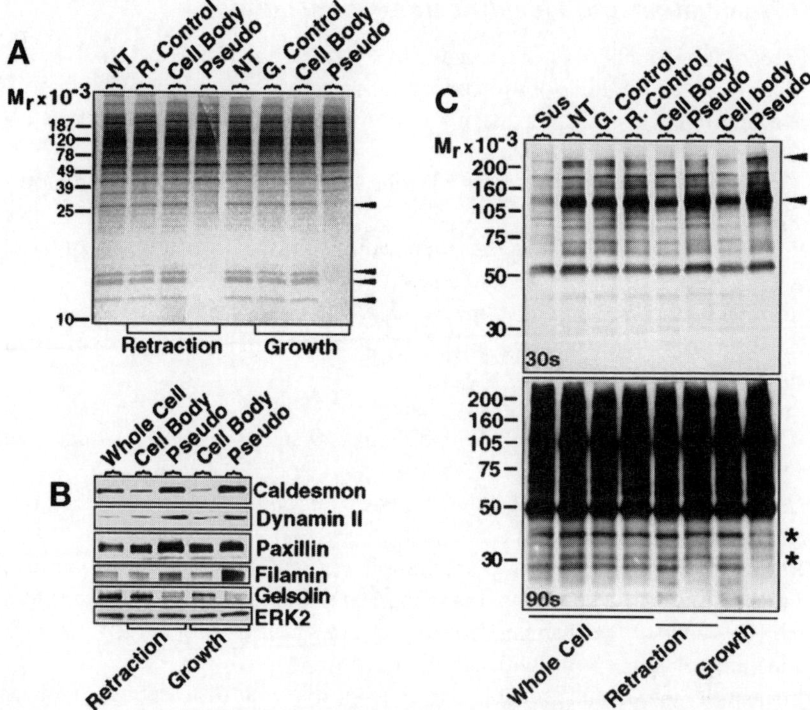

Fig. 3. Biochemical characterization of cytoskeletal-regulatory proteins in growing and retracting pseudopodia. **(A)** NIH 3T3 cells were allowed to extend pseudopodia toward an L-α-lysophosphatidic acid (LPA) gradient (100 ng/mL) for 60 min (growth), or the LPA gradient was removed and pseudopodia allowed to retract for 30 min (retraction). Proteins (10 μg) isolated from the cell body on the upper membrane surface or pseudopodia (Pseudo) on the lower membrane surface were resolved by one-dimensional SDS-PAGE and Coomassie blue staining. Total cellular proteins were also isolated from cells attached to culture dishes either not treated (NT) or treated with a uniform concentration of LPA for 60 min (growth control). Cells were also treated with a uniform concentration of LPA for 60 min and then washed and proteins isolated after 30 min of retraction (retraction control). Arrowheads indicate nuclear histone proteins, which are absent in purified pseudopodia. **(B)** Proteins isolated as described in panel **A** were analyzed by Western blotting using antibodies to the indicated proteins. Whole cell represents total cellular protein isolated from untreated cells attached to fibronectin-coated dishes for 90 min. **(C)** NIH 3T3 cells were either held in suspension for 30 min (sus) or allowed to attach for 2 h to either fibronectin-coated culture dishes or the upper surface of a 3.0-μm porous membrane coated with fibronectin. Whole cells on culture dishes or pseudopodia in the growth and retraction phase were isolated as described in panel **A** and analyzed for tyrosine phosphorylation by Western blotting with anti-phosphotyrosine antibodies. Blots treated with ECL reagent were exposed to film for 30 and 90 s. Arrowheads indicate proteins with increased tyrosine phosphorylation in purified pseudopodia. Asterisk shows proteins with increased tyrosine phosphorylation in retracting pseudopodia. R, control, retraction control; G, control, growth control. (Reprinted with permission from **ref. 3**.)

3.8. Visualization and Quantification of Pseudopodia

1. Prepare the chambers as described above (*see* **Subheadings 3.1.** to **3.4.**).
2. Make a fresh saturated solution of crystal violet stain by adding 125 mg to 25 mL of a solution containing 625 µL of Solution A and 740 µL of Solution B in a 50-mL tube and vortexing vigorously.
3. Filter the solution into a new 50-mL tube using Whatman no. 2 filter paper and a plastic funnel.
4. Immediately before staining the filters, add 500 µL of 100% ethanol to the crystal violet solution to make the final concentration 2%.
5. Add 2 mL of crystal violet stain to each well of a new six-well dish.
6. Remove filters with attached cells from dish and quickly but gently rinse in PBS at room temperature.
7. Place the filters in the crystal violet stain for 15 min.
8. Wash the chamber thoroughly with deionized water. Shake off residual water and place in a clean dry well.
9. Use a cotton swab to wipe off the cell body on top of the filter and then rinse the chamber with water.
10. Repeat **step 8** until cell bodies are completely removed from the top of the filter.
11. To visualize stained pseudopodia using standard microscopy, cut the filter along the outer edge of the chamber using a sharp razor blade. Mount the filter on a glass slide and visualize by brightfield microscopy.
12. To quantify the stained pseudopodia, elute the dye with 10% acetic acid for analysis with an ELISA plate reader. Alternatively, determine the relative amount of pseudopodia protein for each condition using the micro BCA protein kit according to the manufacturer's recommended methodology.
13. Alternatively, to visualize pseudopodia using high-resolution imaging, prepare the chambers as described above (*see* **Subheadings 3.1.** to **3.4.**). Fix filter in 4% paraformaldehyde and process for immunofluorescent microscopy using standard methods. Cut the filter from the chamber, mount it on a glass slide using fluoromount or other mounting media, and visualize using fluorescence microscopy.

4. Notes

1. We have worked extensively with NIH 3T3 and COS-7 cells, but primary and cancer cell lines work, too. Immune cells may also work, but the pore size will have to be reduced so that the cells do not crawl through (~0.4–1.0 µm). It is also important to utilize low-passage cells (<20) because long-term passaging increases basal membrane protrusion and random pseudopodia formation.
2. Starvation of the cells is important to remove growth factors and ECM proteins that stimulate random pseudopodia formation. Each cell type will require a distinct starvation protocol. COS-7 cells are starved in the complete absence of FBS, whereas NIH 3T3 cells require 0.5% FBS in the medium to maintain viability.
3. To dissolve 5 mg human fibronectin, add 5 mL milli Q water filtered with 0.22-µm Tube Top filter (Corning) to the vial and let the fibronectin dissolve for 1 h at

room temperature. Do not vortex the vial because this will cause the proteins to aggregate.

4. 6.5-mm Transwells in 24-well plate format with pore size 3.0 μm can also be used and are available from Corning.

5. Fibronectin is the ECM protein used here, but different ECM proteins, such as collagen or vitronectin, may also be used depending on what cell type one uses.

6. The filter must be checked for leaks around the edge where it attaches to the plastic chamber. We recommend checking for leakiness prior to adding the ECM solution by adding 2 mL of PBS to the lower chamber only. Incubate at room temperature for at least 30 min. If the filter is not completely sealed, the solution will flow to the upper chamber along the outer edge. A properly sealed filter will completely prevent the solution from flowing to the upper chamber.

7. The number of cells used for each filter depends on the type of cells and their size. Empirically, $1.5–1.8 \times 10^6$ cells are optimal for COS-7 and $1.0–1.2 \times 10^6$ cells are optimal for NIH 3T3.

8. Be careful not to disturb the fibronectin coating on the underside of the filter. If the filter must be set down, we recommend placing the filter upside-down on the lid of the six-well plate until cells are ready to be added (**step 7**).

9. The cells must be appropriately attached and spread on the filter to ensure correct sensing of the chemoattractant or ECM gradient.

10. LPA is the chemokine used here, but different chemoattractants, such as 20 ng/mL platelet-derived growth factor, can also be used to stimulate growth of pseudopodia effectively.

11. Setting up the chemoattractant gradient correctly is the key to successfully stimulating pseudopodia growth. The filters need to be prealigned before being placed into the chamber containing LPA so as not to require adjustment after placement. When the dishes are transferred to the growth chamber, they must be carried horizontally and without shaking.

12. Extension of pseudopodia through 3.0-μm pores can be detected 10–15 min after exposure to the LPA gradient and proceeds linearly for 60–90 min *(3)*. Formation of the LPA gradient can be steep and linear for at least 3 h. The optimal time for pseudopodia extension will have to be determined for each cell type used; 60 min is optimal for NIH 3T3 and COS-7 cells.

13. The fixation step may be omitted for applications requiring nondenaturing conditions, such as GTPase activity and immunoprecipitation assays *(3)*.

14. 1% SDS will completely solubilize all pseudopodial proteins. However, mild detergents, such as Triton X-100 and NP-40, can be used for other applications, including immunoprecipitation of specific proteins or protein complexes *(3)*.

15. A broken membrane can fold over on itself and trap lysate or cause the loss of pseudopodia.

16. Do not break the parafilm or the lysate will be trapped between the glass plate and parafilm.

17. Boiling the lysate in 1% SDS will denature all proteins and inactivate enzymes. If the lysates are used for applications requiring nondenaturing conditions, the boiling step can be omitted.

Acknowledgments

We thank R. Hanley for her helpful advice and comments on this manuscript. This work is supported by National Institutes of Health Grant GM68487 (to R.L.K.), CA97022 (to R.L.K.), and a grant from the Susan G. Komen Breast Cancer Foundation (to Y.W.).

References

1. Lauffenburger, D. A. and Horwitz, A. F. (1996) Cell migration: a physically integrated molecular process. *Cell* **84,** 359–369.
2. Parent, C. A. and Devreotes, P. N. (1999) A cell's sense of direction. *Science* **284,** 765–770.
3. Cho, S. Y. and Klemke, R. L. (2002) Purification of pseudopodia from polarized cells reveals redistribution and activation of Rac through assembly of a CAS/Crk scaffold. *J. Cell Biol.* **156,** 725–736.
4. Brahmbhatt, A. A. and Klemke, R. L. (2003) ERK and RhoA differentially regulate pseudopodia growth and retraction during chemotaxis. *J. Biol. Chem.* **278,** 13,016–13,025.
5. Jin, T., Zhang, N., Long, Y., Parent, C. A., and Devreotes, P. N. (2000) Localization of the G protein betagamma complex in living cells during chemotaxis. *Science* **287,** 1034–1036.
6. Meili, R., Ellsworth, C., Lee, S., Reddy, T. B., Ma, H., and Firtel, R. A. (1999) Chemoattractant-mediated transient activation and membrane localization of Akt/PKB is required for efficient chemotaxis to cAMP in Dictyostelium. *EMBO J.* **18,** 2092–2105.
7. Parent, C. A., Blacklock, B. J., Froehlich, W. M., Murphy, D. B., and Devreotes, P. N. (1998) G protein signaling events are activated at the leading edge of chemotactic cells. *Cell* **95,** 81–91.
8. Servant, G., Weiner, O. D., Herzmark, P., Balla, T., Sedat, J. W., and Bourne, H. R. (2000) Polarization of chemoattractant receptor signaling during neutrophil chemotaxis. *Science* **287,** 1037–1040.
9. Chodniewicz, D. and Klemke, R. L. (2004) Guiding cell migration through directed extension and stabilization of pseudopodia. *Exp. Cell Res.* **301,** 31–37.
10. Pollard, T. D. and Borisy, G. G. (2003) Cellular motility driven by assembly and disassembly of actin filaments. *Cell* **112,** 453–465.
11. Ridley, A. J., Schwartz, M. A., Burridge, K., et al. (2003) Cell migration: integrating signals from front to back. *Science* **302,** 1704–1709.
12. Ge, L., Ly, Y., Hollenberg, M., and DeFea, K. (2003) A beta-arrestin-dependent scaffold is associated with prolonged MAPK activation in pseudopodia during protease-activated receptor-2-induced chemotaxis. *J. Biol. Chem.* **278,** 34,418–34,426.
13. Lin, Y. H., Park, Z. Y., Lin, D., et al. (2004) Regulation of cell migration and survival by focal adhesion targeting of Lasp-1. *J. Cell. Biol.* **165,** 421–432.
14. Wozniak, M. A., Kwong, L., Chodniewicz, D., Klemke, R. L., and Keely, P. J. (2005) R-Ras controls membrane protrusion and cell migration through the spatial regulation of Rac and Rho. *Mol. Biol. Cell.* **16,** 84–96.

6

Dynamic Assessment
of Cell–Matrix Mechanical Interactions
in Three-Dimensional Culture

W. Matthew Petroll

Summary

Cell–matrix mechanical interactions play a defining role in a range of biological processes such as developmental morphogenesis and wound healing. Despite current agreement that fibroblasts exert mechanical forces on the extracellular matrix (ECM) to promote structural organization of the collagen architecture, the underlying mechanisms of force generation and transduction to the ECM are not completely understood. Investigation of these processes has been limited, in part, by the technical challenges associated with simultaneous imaging of cell activity and fibrillar collagen organization. To overcome these limitations, we have developed an experimental model in which cells expressing proteins tagged with enhanced green fluorescent protein are plated inside fibrillar collagen matrices, and high magnification time-lapse differential interference contrast and fluorescent imaging is then performed. Using this system, focal adhesion movement and reorganization in isolated cells can be directly correlated with collagen matrix deformation and changes in the mechanical behavior of fibroblasts can be assessed over time.

Key Words: Focal adhesions; focal complexes; cell motility; collagen matrices; actomyosin; cell mechanics; three-dimensional culture.

1. Introduction

Historically, our ability to investigate cell mechanical behavior has been limited, in part, by the technical challenges associated with simultaneous imaging of cell activity and measurement of cellular forces. Most previous studies of cell mechanical behavior have used planar elastic substrates *(1)*. In early studies, silicon substrates were used to estimate the forces generated by migrating cells *(2–7)*. More recently, Wang and coworkers have used flexible polyacrylamide

From: *Methods in Molecular Biology, vol. 370:*
Adhesion Protein Protocols, Second Edition
Edited by: A. S. Coutts © Humana Press Inc., Totowa, NJ

sheets embedded with fluorescent microspheres to study various aspects of cell mechanical behavior *(8–14)*. This elegant model has provided important insights into many aspects of cell mechanical behavior. However, cells reside within complex three-dimensional (3D) extracellular matrices in vivo, and extracellular matrix (ECM) geometry has been shown to effect cell morphology and adhesion organization and mechanical activity *(15–21)*.

An alternative to planar elastic substrates is the fibroblast populated collagen lattice model, in which cells are plated inside a 3D fibrillar collagen matrix *(4,22–24)*. The fibroblast populated collagen lattice is a standard in vitro model for studying the mechanisms regulating cell-mediated matrix reorganization and wound contraction *(25)*. Measuring overall collagen matrix contraction and/or force generation by fibroblasts using this model has provided valuable insights into the signaling pathways involved in various aspects of cell–matrix mechanical interactions *(25–39)*. However, these global measurements do not provide a detailed understanding of the underlying pattern of force generation and matrix reorganization at the cellular and subcellular level.

We have developed a new experimental model for dynamic investigation of cell–matrix mechanical interactions by plating cells transfected to express enhanced green fluorescent protein (EGFP)-tagged proteins inside fibrillar collagen matrices and performing high-magnification time-lapse differential interference contrast (DIC) and fluorescent imaging. Using this system, focal adhesion movement and reorganization in isolated cells can be directly correlated with collagen matrix deformation, and changes in the mechanical behavior of fibroblasts in response to both biochemical and biophysical signals can be assessed in real time *(40–43)*.

2. Materials

2.1. Cell Culture and Transfection

1. Complete media consisting of Dulbecco's modified Eagle's medium (DMEM; GIBCO Invitrogen cell culture, Carlsbad, CA) supplemented with 1% penicillin, 1% streptomycin, and 1% Fungizone (Biowhittaker, Inc., Wakersville, MD) and 10% fetal bovine serum (Sigma Chemical Corp., St. Louis, MO).
2. Solution of trypsin (0.05%) and ethylenediamine tetraacetic acid (EDTA) (0.53 m*M*) from Gibco/BRL.
3. To label focal adhesions in living cells, human zyxin in a pEGFP-N1 vector *(13, 44–46)*. This probe was a generous gift of Professor J. Wehland and coworkers (BGF, Braunschweig, Germany).
4. The unmodified pEGFP-N1 expression vector (Clontech Laboratories, Inc., Mountain View, CA).
5. Lipofectamine PLUS (Invitrogen, Carlsbad, CA).

2.2. Preparation of Fibroblast-Populated Collagen Matrices

1. Bovine dermal collagen (PureCol, Inamed Corp., Fremont, CA), DMEM, and 1 *N* NaOH (Sigma).
2. Delta T culture dishes (Bioptechs, Inc., Butler, PA).

2.3. Time-Lapse Digital Imaging

1. A Nikon TE300 inverted microscope with fluorescence and DIC imaging modules (Nikon Inc., Melville, NY), a remote focus accessory (which controls microscope stage vertical movement), two high-speed filter wheels for rapid selection of excitation and emission filters and shuttering of epifluorescent illumination (Lambda 10-2, Sutter Instruments Company, Novato, CA), a transmitted light shutter (Uniblitz, Vincent Associates, Rochester, NY) and a high-resolution cooled charge-coupled device digital camera (CoolSnap HQ, Roper Scientific, Tuscan, AZ). A personal computer running MetaView (Molecular Devices, Sunnyvale, CA) is used to control filter wheels, shuttering, focus, and digital image acquisition.
2. A Delta T temperature controller and stage adapter, objective heater and controller, and microperfusion pump (Bioptechs, Inc).
3. Perfusion media: complete media containing 50 m*M* *N*-2-hydroxyethylpiperazine-*N'*-2-ethanesulfonic acid (HEPES) buffer (Sigma).

3. Methods

1. Corneal fibroblasts are maintained in 25-cm^2 tissue culture flasks (Costar, Cambridge, MA) using complete media (*see* **Note 1**).
2. Fibroblasts are passaged using trypsin–EDTA onto six-well plates (40,000 cells/well) using complete media without antibiotics 24 h before transfection.
3. Each EGFP expression vector is first expressed in *Escherichia coli* DH5α (Life Technologies, Rockville, MD), and clones expressing the plasmid are selected based on resistance to kanamycin (30 μg/mL). The plasmids are then further purified using the Qiagen plasmid mini kit (Qiagen Inc., Valencia, CA). The purity of the plasmid is confirmed from the size of the cleavage products after digestion with appropriate restriction enzymes.
4. For each well, 0.5 μg of purified GFP–zyxin DNA in 100 μL DMEM is precomplexed with 3 μL PLUS reagent for 15 min. Meanwhile, 2.5 μL of lipofectamine reagent is mixed with 100 μL of DMEM. Precomplexed DNA and diluted lipofectamine reagent are then mixed and incubated for 15 min at room temperature.
5. Complete media is removed from each well and replaced with the DNA + lipofectamine reagent complex mixed with an additional 800 μL of DMEM (total of 1 mL of media per well). Cells are then placed in a humidified incubator (37°C, 5% CO_2) for 3 h, after which 1 mL of DMEM + 20% fetal bovine serum is added to each well (for a total of 2 mL of media per well).
6. After an additional 24 h of incubation, transfection media is replaced with complete media (*see* **Note 2**).

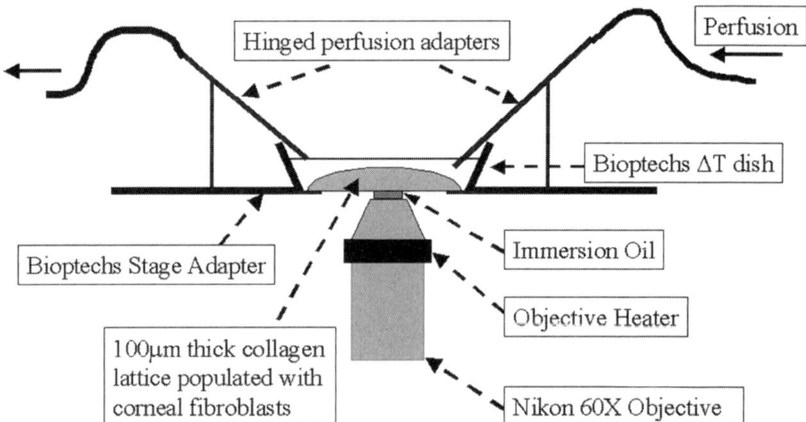

Fig. 1. Schematic of experimental setup on the microscope for time-lapse imaging. (Reprinted with permission from **ref. 56**.)

3.2. Preparation of Fibroblast-Populated Collagen Matrices

1. Hydrated collagen matrices are prepared by mixing 400 μL of bovine dermal collagen with 50 μL of 10X DMEM and 50 μL of NaOH to achieve a final collagen concentration of 2.48 mg/mL. The collagen solution is kept on ice at all times. Additional NaOH is added as needed to achieve a final pH of 7.2–7.4.

2. GFP–zyxin transfected or control (pEGFP-N1 transfected or untransfected) corneal fibroblasts are harvested with trypsin–EDTA (*see* **Note 3**). The cells are diluted using complete media to a final concentration of 1.7×10^5 cells/mL. Fifty microliters of cell suspension is mixed with 450 μL of collagen solution. The cell–collagen mixture is preincubated at 37°C for 5 min, and 30-μL aliquots (containing approx 500 cells) are then poured onto Delta T culture dishes (Bioptechs, Inc.). Each aliquot is spread over a central 12-mm-diameter circular region on the dish and is approx 100 μm thick in the center following polymerization. The dish is then placed in a humidified incubator (37°C, 5% CO_2) for 60 min for polymerization, overlaid with 1.5 mL of complete media (*see* **Note 4**).

3.3. Time-Lapse Digital Imaging

1. The experimental setup for time-lapse imaging is shown in **Fig. 1**. With the Bioptechs system, cells are both cultured and imaged in special dishes with glass bottoms of no. 1 cover slip thickness (~0.17 mm); this increases the available working distance as compared to plastic and is also compatible with DIC imaging. A Delta T culture dish containing a collagen matrix is removed from the incubator and clipped onto the special stage adapter. Contacts on the stage adapter deliver electrical current across a special resistive coating on the bottom of the Delta T dish. A thermister on the stage adapter contacts the bottom of the dish and monitors the dish temperature. The stage adapter is connected to a temperature control box that

adjusts the current to maintain a desired set point through a feedback loop. Because an immersion objective is used, the objective is also heated using a collar interfaced to a similar control box. The dish is continuously perfused with fresh media containing HEPES (bubbled with 5% CO_2) while on the microscope stage at a rate of 6 mL/h using a Bioptechs microperfusion pump (*see* **Note 5**).

2. After allowing the cells to acclimate to the microincubation system for 1 h, the activity of a single cell is imaged for up to 5 h using a Nikon 60X oil immersion objective (1.4 NA, 210 µm free working distance) (*see* **Note 6**). Isolated cells (no neighbors within 300 µm) are selected to minimize mechanical interference from the activity of other fibroblasts within the matrix. In our model, cells are generally imaged approx 18 h after seeding on the gel; this allows them to develop a more consistent bipolar spindle-shaped morphology as is observed during in vivo corneal wound healing *(17,47–49)*. Images are automatically acquired at 1- to 3-min intervals using MetaView. Because all of the focal adhesions are not in a single plane, 3D datasets are obtained at each time point by repeating the acquisition at four to six sequential focal planes in z steps of 1–3 µm. For each time point two images are acquired in rapid succession at each focal plane: one using bandpass fluoroisothiocyanate (FITC) excitation and emission filters (for EGFP), and one using DIC. A DAPI/FITC/PI beam splitter is used for all acquisitions (Chroma Technology Corp, Rockingham, VT) (*see* **Note 7**).

3.4. Image Processing

1. Eighteen hours after plating inside 3D collagen matrix, corneal fibroblasts generally have a bipolar morphology with thin pseudopodial processes and do not form broad lamellipodia (**Fig. 2**). Cells are always aligned nearly parallel to the dish on which the collagen matrix was plated. DIC imaging allows detailed visualization of the cells and the fibrillar collagen surrounding them, and different populations of collagen fibrils are generally observed in each image of the 3D DIC datasets (*see* **Fig. 2A,B,C**). Individual collagen fibrils are easily discerned adjacent to cells (**Fig. 2**, arrows), and also crossing above (**Fig. 2C**, arrowheads) and below (**Fig. 2A**, arrowheads) the cell body. In our cells, GFP–zyxin is organized into adhesions that are most concentrated along the pseudopodia at the leading edge of cells, although adhesions are also scattered along the cell body (**Fig. 3B,C**). Diffuse background GFP–zyxin labeling is also present, suggesting a cytoplasmic pool of this protein. Counterstaining with vinculin has been used to confirm that GFP–zyxin is properly organized into focal adhesions *(41)*. Control cells transfected with the EGFP-N1 vector alone had bright and diffuse nonspecific cytoplasmic and nuclear labeling, with no specific labeling of adhesions (not shown).

2. Because all of the focal adhesions are not generally visible in a single plane, 3D reconstructions of the z-series are generated. For each time interval, 3D reconstructions of GFP–zyxin are generated by applying "flatten background" operation to each 2D image in the z-series and then combining the images using a maximum intensity projection with MetaMorph. Overlaying of GFP–zyxin maximum intensity projections (**Fig. 3E**) and individual DIC images (**Fig. 3A**) using the "Overlay

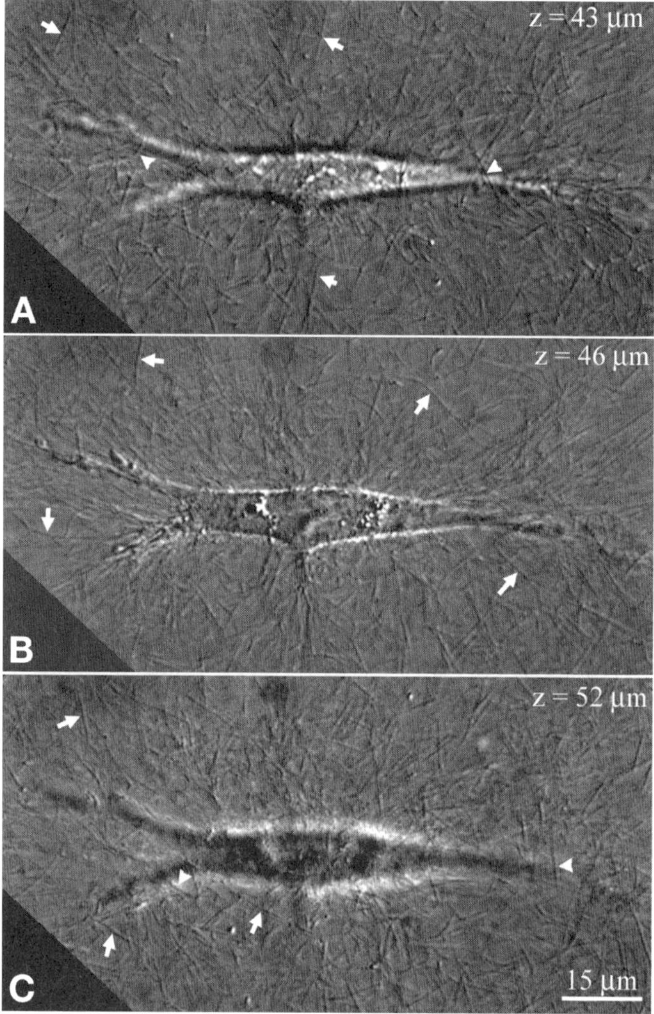

Fig. 2. Three differential interference contrast images from a five-image z-series of an untransfected corneal fibroblast plated inside a three-dimesional collagen matrix. The z position shown is relative to the bottom of the collagen matrix. Individual collagen fibrils are easily discerned adjacent to the cell (arrows), and also crossing above (**C**, arrowheads) and below (**A**, arrowheads) the cell body. A different population of collagen fibrils is observed in each image. (Reprinted with permission from **ref. *40*.**)

Images" function in MetaMorph allows the organization of focal adhesions and collagen fibrils to be directly compared (**Fig. 3F**). Time-lapse movies can be generated by performing the overlay for the entire sequence of images. Color-coded reconstructions can also be generated by displaying the bottom planes in red and the top planes in green (**Fig. 3D**) (*see* **Note 8**).

3. Using high-magnification DIC imaging, movement and realignment of collagen fibrils during cell migration can be directly observed. In order to quantify the ECM deformation, the *x,y* coordinates of landmarks in DIC images are determined using the "track points" feature in MetaMorph (*see* **Note 9**). In order to measure the location of adhesions over time, we use the Integrated Morphometry Analysis (IMA) feature of MetaMorph. The location, intensity, and morphometry of each above-threshold adhesion can then be recorded for frames of interest in the sequence using IMA (*see* **Note 10**). In order to display the ECM and adhesion displacements, we developed a custom program using Visual Basic (Microsoft, Redmond, WA). This program calculates the displacement vectors from the measured *x,y* coordinates and generates an image with arrows of corresponding magnitude and direction (*see* **Note 11**).

4. Notes

1. We have performed transfections on both primary cultures of rabbit corneal fibroblasts and immortalized corneal fibroblasts. For primary cultures, the amount of mechanical activity generally decreases with the number of passages. This decrease can be identified using standard collagen matrix contraction assays. For corneal fibroblasts, all experiments are performed within the first three passages, during which the mechanical activity remains stable.

2. The EGFP-N1 vector has a neo selection cassette; thus, when using immortalized cell populations, stably transfected cells can be selected by using complete media containing G418. EGFP-expressing cells can then be further purified using fluorescence activated cell sorting. In addition to EGFP–zyxin, we have also used EGFP–α-actinin, which labels both focal adhesions and stress fibers (*42,50*). Of course, all of the transfection procedures described should be optimized empirically for other cell types and/or different GFP expression vectors.

3. We typically plate cells inside matrices 24–48 h after transient transfection. This gives them adequate time to recover from the stress of the transfection procedure while still maintaining a high level of expression. It is important that the cells are subconfluent prior to plating inside the collagen matrices (60–80%), because confluent cells typically exhibit less mechanical activity.

4. Collagen matrices are not as adherent to the glass Bioptechs dishes as they are to plastic. Thus, extreme care should be taken when overlaying media onto the matrices to avoid detachment. It is possible to re-use the Delta T dishes after soaking in alcohol and exposing them to ultraviolet light. However, collagen matrices tend to adhere better to new dishes. Unfortunately, we have not found precoating the dishes with collagen to be helpful in preventing occasional detachment.

5. The temperature set points for the dish and objective controllers should be determined empirically for a given system. For calibration experiments, temperature-sensitive disks (from Bioptechs) that change color at 37°C can be placed in the dish media or on the objective immersion fluid. We have found it best to use a set point 1–3°C lower than that which causes the disks to change color, since in our experience cells are more tolerant of temperatures below 37°C than they are of higher

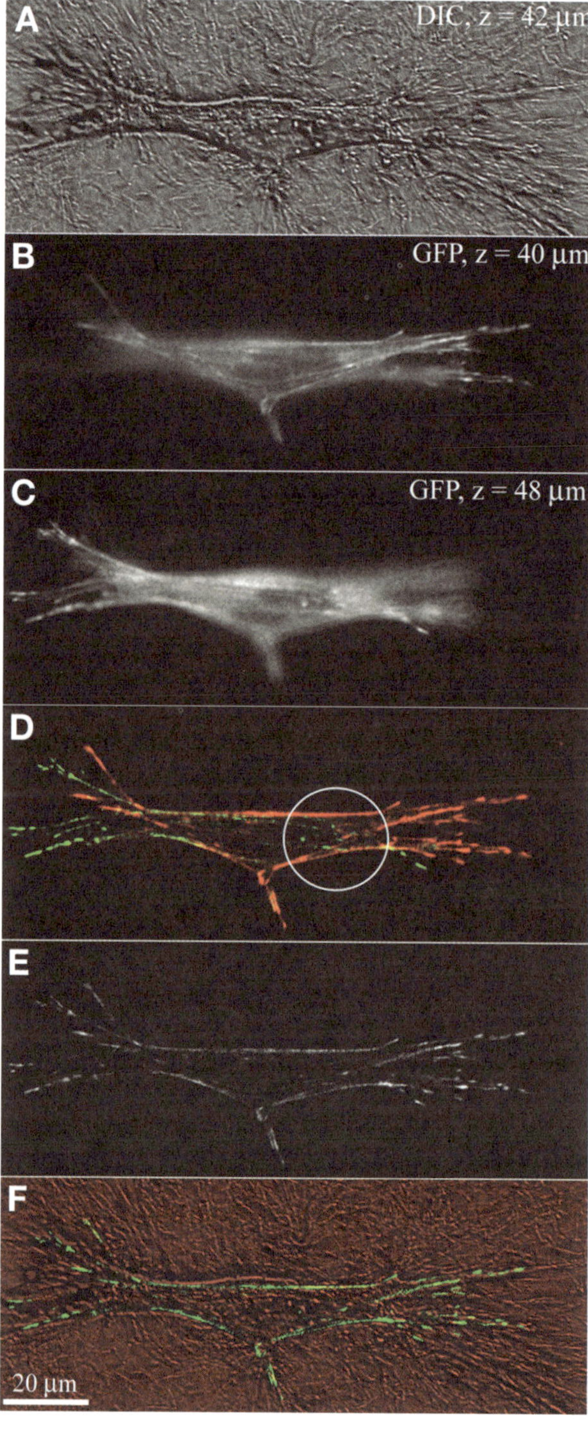

temperatures. Note that temperature-control systems from other vendors can be used as long as they support the use of dishes with glass bottoms of cover slip thickness, which are required for high-magnification imaging. An "open" perfusion system is used instead of a chamber because it facilitates the direct addition of reagents to the culture dish (e.g., cytochalasin D) and also allows microneedles to be inserted into the matrix to alter local mechanical properties *(43)*.

6. Both 40X and 60X oil immersion lenses provide adequate resolution to distinguish collagen fibrils using DIC. However, the 60X is needed to distinguish GFP-tagged focal adhesions within the collagen matrices. Our 100-μm-thick matrices are well within the 210 μm working distance of the Nikon 60X objective. Although there is a refractive index mismatch between oil and the aqueous media within the collagen matrices, we have found empirically that oil immersion lenses provide higher quality images than water immersion lenses in our model. Because oil does not evaporate, it is also better suited for time-lapse imaging over several hours.

7. When performing fluorescent imaging, the effects of photobleaching and phototoxicity need to be considered. In general, the amount of exposure to the fluorescent excitation (mercury arc lamp) should be minimized. In our experiments, neutral density filters (40 or 80% transmittance) are used, and exposure times are kept to a minimum by using 2×2 on-chip camera binning. With binning, the images are 696×520 pixels, and the typical camera exposure time for each EGFP image is 1 s. Although costly, it is important to use a highly sensitive cooled charge-coupled device camera, because it provides high signal-to-noise ratio with short exposure times. Also, the time interval used and the number of planes collected at each time point should be reduced if necessary to prevent phototoxicity. This can be determined empirically by comparing the mechanical activity of cells (from DIC images) with and without fluorescent imaging of EGFP. Because the z-axis position of cells may drift somewhat over time, it is useful to "bracket" the cell when collecting 3D stacks such that at least one plane is collected slightly above and below the cell's initial position Note that all shutters are closed between acquisitions.

Fig. 3. *(Opposite page)* Image processing and display of three-dimensional (3D) datasets. Differential interference contrast (DIC) (**A**), green fluorescent protein (GFP) (**B–E**), and combined (**F**) images of a corneal fibroblast expressing GFP–zyxin are shown. (**A**) A single DIC image showing the fibrillar collagen organization surrounding the cell. (**B,C**) The top and bottom GFP images from a five-image z-series. GFP–zyxin was organized into adhesions that were most concentrated along pseudopodia at the ends of the cell. (**D**) Color-coded reconstruction produced by performing a flatten background operation on each image in the z-series, then overlaying the bottom three images in red, and the top two images in green. Adhesions are visualized on both the ventral (red) and dorsal (green) surface of the cell body in some regions (circle). (**E**) 3D reconstruction produced using a maximum intensity projection. (**F**) Color overlay of **E** (green) and **A** (red) allows focal adhesion and collagen organization to be directly compared. (Reprinted with permission from **ref. 40**.)

8. 3D reconstructions of the DIC images is difficult because of the complex content of the images; thus 2D DIC images from the same focal plane as the adhesions of most interest are used for the color overlays. As elegantly demonstrated by Friedl and coworkers, reflected light laser scanning confocal microscopy (LSCM) can be used to visualize the true 3D structural organization of the fibrillar collagen *(19,51–55)*. In our laboratory, LSCM of living corneal fibroblasts transfected to express GFP–zyxin or GFP–α-actinin has been used to correlate matrix reorganization with focal adhesion and/or stress fiber organization *(56)*. Although better suited for studying ECM remodeling, LSCM is much slower and more phototoxic than wide field microscopy and thus is not as useful for studying rapid changes in focal adhesion organization and ECM deformation.

9. In earlier studies, fluorescent beads were incorporated into the matrix to allow tracking of ECM deformation *(41)*. However, we have established that tracking landmarks on collagen fibrils in the DIC images provides an accurate measure of ECM deformation *(41)*. The ability to track ECM deformation directly (i.e., without using beads) is important, since beads degrade the overall quality of the DIC images, use a fluorescent wavelength band that could potentially be used for a second fluorescent label, and are sometimes internalized by the cells via endocytosis. Finally, tracking ECM deformation directly from the DIC images provides a much more uniform map of the ECM deformation around the cell.

10. IMA requires that the focal adhesions can be segmented from the rest of the image using an intensity threshold. To achieve this, preprocessing includes the "flatten background" and "equalize stack" procedures, which generally allow a single threshold to be used for the entire time-lapse sequence. Although somewhat time-consuming, IMA is useful because it provides a semi-automated assessment of the morphometry and intensity of a structure of interest in addition to its position (centroid).

11. The dynamic interactions between focal adhesions and collagen fibrils are best appreciated by generating time-lapse color overlay movies in Metamorph (*see* ref. *40* for movie links). However, tracking the movement of the focal adhesions and ECM allows their displacements to be compared on still images. As shown in **Fig. 4**, during pseudopodial extension, new focal adhesions tend to form in a linear pattern along collagen fibrils (arrowheads in **Figs. 4A,B**), whereas existing adhesions move backward. This results in pulling in of the ECM in front of the cells (blue vectors in **Fig. 4B,C**). Meanwhile, adhesions at the base of pseudopodia move toward those at the tip, resulting in contractile-like shortening and ECM compression (green vectors in **Fig. 4C**). The corresponding compression of collagen fibrils (via bending) can be directly visualized in many cases (*see* **Figs. 4A,B,** double arrows). Although the methods are beyond the scope of this review, the map of ECM displacements can also be used to generate quantitative ECM strain maps using finite element modeling, as previously described *(42,43)*.

Acknowledgments

Supported by a grant from the National Institutes of Health (EY13322) and a Lew R. Wasserman Merit award and departmental grant from Research to Prevent Blindness.

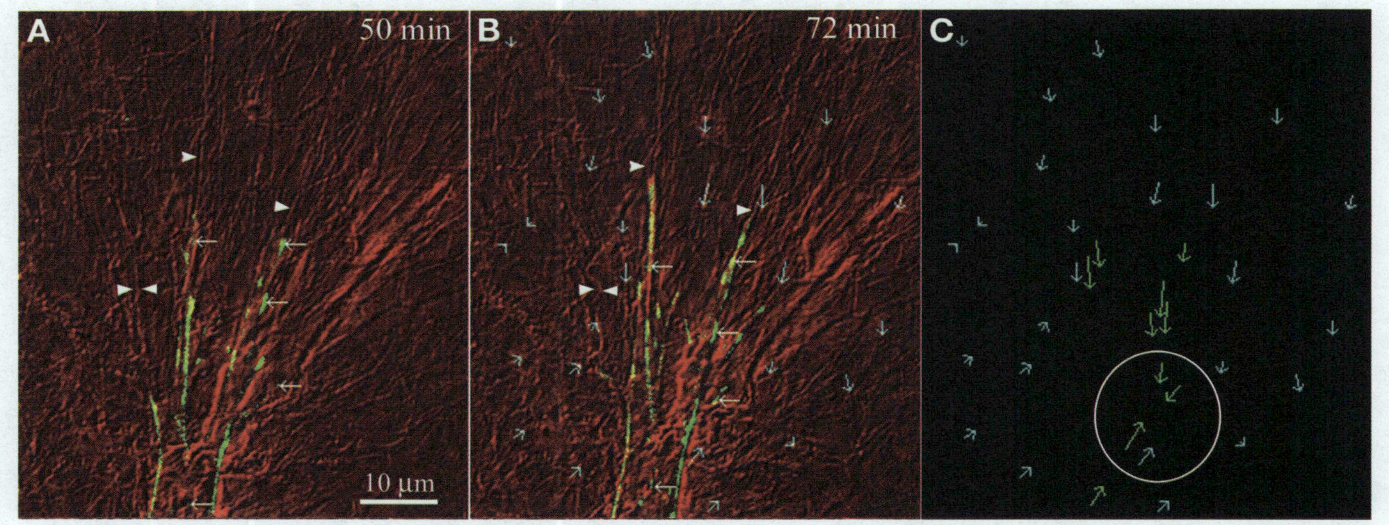

Fig. 4. Summary of cell–matrix interactions during extension of pseudopodia. (**A,B**) Green fluorescent protein/differential interference contrast color overlays at different time points; the position of corresponding adhesions are denoted by white arrows. (**C**) Vectors denoting the magnitude and direction of matrix (blue arrows) and adhesion (green arrows) displacements. New focal adhesions appear to form along collagen fibrils (arrowheads) at the front of the pseudopodia, whereas existing adhesions move backward (compare position of white arrows in **A** and **B**). This results in pulling in of the extracellular matrix (ECM) above the cell (blue arrows, **B** and **C**). Adhesions at the base of the pseudopodia move toward those at the tip, resulting in contractile-like shortening and ECM compression (circled region in **C**). A corresponding compression of collagen fibrils (via bending) could be directly visualized (double arrowheads). Overall, the pattern of ECM and focal adhesion movement was similar (**C**). (Reprinted with permission from **ref. *40***.)

References

1. Harris, A. K., Wild, P., and Stopak, D. (1980) Silicone rubber substrata: a new wrinkle in the study of cell locomotion. *Science* **208,** 177–189.
2. Harris, A. K., Stopak, D., and Wild, P. (1981) Fibroblast traction as a mechanism for collagen morphogenesis. *Nature* **290,** 249–251.
3. Harris, A. K. (1986) Cell traction in relationship to morphogenesis and malignancy. *Dev. Biol.* **3,** 339–357.
4. Stopak, D. and Harris, A. K. (1982) Connective tissue morphogenesis by fibroblast traction. *Dev. Biol.* **90,** 383–398.
5. Lee, J., Leonard, M., Oliver, T., Ishihara, A., and Jacobson, K. (1994) Traction forces generated by locomoting keratocytes. *J. Cell Biol.* **127,** 1957–1964.
6. Oliver, T., Dembo, M., and Jacobson, K. (1995) Traction forces in locomoting cells. *Cell Motil. Cytoskeleton* **31,** 225–240.
7. Balaban, N. Q., Schwarz, U. S., Riveline, D., et al. (2001) Force and focal adhesion assembly: a close relationship studied using elastic micropatterned substrates. *Nat. Cell Biol.* **3,** 466–472.
8. Pelham, R. J. Jr. and Wang, Y. (1997) Cell locomotion and focal adhesions are regulated by substrate flexibility. *Proc. Natl. Acad. Sci. USA* **94,** 13,661–13,665.
9. Pelham, R. J. and Wang, Y. (1999) High resolution detection of mechanical forces exerted by locomoting fibroblasts on the substrate. *Mol. Biol. Cell* **10,** 935–945.
10. Wang, Y. L. and Pelham, R. J. Jr. (1998) Preparation of a flexible, porous polyacrylamide substrate for mechanical studies of cultured cells. *Methods Enzymol.* **298,** 489–496.
11. Munevar, S., Wang, Y., and Dembo, M. (2001) Traction force microscopy of migrating normal and H-ras transformed 3T3 fibroblasts. *Biophys. J.* **80,** 1744–1757.
12. Pelham, R. J. Jr. and Wang, Y. L. (1998) Cell locomotion and focal adhesions are regulated by the mechanical properties of the substrate. *Biol. Bull.* **194,** 348–350.
13. Beningo, K. A., Dembo, M., Kaverina, I., Small, J. V., and Wang, Y. L. (2001) Nascent focal adhesions are responsible for the generation of strong propulsive forces in migrating fibroblasts. *J. Cell Biol.* **153,** 881–888.
14. Beningo, K. A., Dembo, M., and Wang, Y. L. (2004) Responses of fibroblasts to anchorage of dorsal extracellular matrix receptors. *Proc. Natl. Acad. Sci.* **101,** 18,024–18,029.
15. Bard, J. B. L. and Hay, E. D. (1975) The behavior of fibroblasts from the developing avian cornea: morphology and movement *in situ* and in vitro. *J. Cell Biol.* **67,** 400–418.
16. Cukierman, E., Pankov, R., and Yamada, K. M. (2002) Cell interactions with three-dimensional matrices. *Curr. Opin. Cell Biol.* **14,** 633–639.
17. Cukierman, E., Pankov, R., Stevens, D. R., and Yamada, K. M. (2001) Taking cell-matrix adhesions to the third dimension. *Science* **294,** 1708–1712.
18. Doane, K. J. and Birk, D. E. (1991) Fibroblasts retain their tissue phenotype when grown in three-dimensional collagen gels. *Exp. Cell Res.* **195,** 432–442.
19. Friedl, P. and Brocker, E.-B. (2000) The biology of cell locomotion within three-dimensional extracellular matrix. *Cell. Mol. Life Sci.* **57,** 41–64.

20. Tomasek, J. J., Hay, E. D., and Fujiwara, K. (1982) Collagen modulates cell shape and cytoskeleton of embryonic corneal and fibroma fibroblasts: distribution of actin, α-actinin and myosin. *Dev. Biol.* **92,** 107–122.
21. Abbott, A. (2003) Biology's new dimension. *Nature* **424,** 870–872.
22. Bell, E., Ivarsson, B., and Merril, C. (1979) Production of a tissue-like structure by contraction of collagen lattices by human fibroblasts of different proliferative potential in vivo. *Proc. Natl. Acad. Sci. USA* **76,** 1274–1278.
23. Elsdale, T. and Bard, J. (1972) Collagen substrata for studies on cell behavior. *J. Cell Biol.* **54,** 626–637.
24. Grinnell, F. and Lamke, C. R. (1984) Reorganization of hydrated collagen lattices by human skin fibroblasts. *J. Cell Sci.* **66,** 51–63.
25. Grinnell, F. (2000) Fibroblast-collagen matrix contraction: growth-factor signalling and mechanical loading. *Trends Cell Biol.* **10,** 362–365.
26. Cheema, U., Yang, S.-Y., Mudera, V., Goldspink, G. G., and Brown, R. A. (2003) 3-D in vitro model of early skeletal muscle development. *Cell Motil. Cytoskeleton* **54,** 226–236.
27. Eastwood, M., McGrouther, D. A., and Brown, R. A. (1994) A culture force monitor for measurement of contraction forces generated in human dermal fibroblast cultures: evidence for cell matrix mechanical signalling. *Biochim. Biophys. Acta* **1201,** 186–192.
28. Kolodney, M. S. and Elson, E. L. (1993) Correlation of myosin light chain phosphorylation with isometric contraction of fibroblasts. *J. Biol. Chem.* **268,** 23,850–23,855.
29. Wakatsuki, T. and Elson, E. L. (2003) Reciprocal interactions between cells and extracellular matrix during remodeling of tissue constructs. *Biophys. Chem.* **100,** 593–605.
30. Arora, P. D., Narani, N., and McCulloch, C. A. G. (1999) The compliance of collagen gels regulates transforming growth factor-beta induction of alpha-smooth muscle actin in fibroblasts. *Am. J. Pathol.* **154,** 871–882.
31. Brown, R. A., Prajapati, R., McGrouther, D. A., Yannas, I. V., and Eastwood, M. (1998) Tensional homeostasis in dermal fibroblasts: mechanical responses to mechanical loading in three-dimensional substrates. *J. Cell. Physiol.* **175,** 323–332.
32. Freyman, T. M., Yannas, I. V., Yokoo, R., and Gibson, L. J. (2002) Fibroblast contractile force is independent of the stiffness which resists the contraction. *Exp. Cell Res.* **272,** 153–162.
33. Parizi, M., Howard, E. W., and Tomasek, J. J. (2000) Regulation of LPA-promoted myofibroblast contraction: role of rho, myosin light chain kinase, and myosin light chain phosphotase. *Exp. Cell Res.* **254,** 210–220.
34. Rosenfeldt, H., Lee, D. J., and Grinnell, F. (1998) Increased c-fos mRNA expression by human fibroblasts contracting stressed collagen matrices. *Mol. Cell. Biol.* **18,** 2659–2667.
35. Shreiber, D. I., Enever, P. A. J., and Tranquillo, R. T. (2001) Effects of PDGF-BB on rat dermal fibroblast behavior in mechnically stressed and unstressed collagen and fibrin gels. *Exp. Cell Res.* **266,** 155–166.

36. Skuta, G., Ho, C.-H., and Grinnell, F. (1999) Increased myosin light chain phosphorylation is not required for growth factor stimulation of collagen matrix contraction. *J. Biol. Chem.* **274,** 30,163–30,168.
37. Vaughan, M. B., Howard, E. W., and Tomasek, J. J. (2000) Transforming growth factor-β1 promotes the morphological and functional differentiation of the myofibroblast. *Exp. Cell Res.* **257,** 180–189.
38. Grinnell, F., Ho, C.-H., Tamariz, E., Lee, D. J., and Skuta, G. (2003) Dendritic fibroblasts in three-dimensional collagen matrices. *Mol. Cell. Biol.* **14,** 384–395.
39. Abe, M., Ho, C.-H., Kamm, K. E., and Grinnell, F. (2003) Different molecular motors mediate platelet-derived growth factor and lysophosphatidic acid-mediated floating collagen matrix contraction. *J. Biol. Chem.* **278,** 47,707–47,712.
40. Petroll, W. M. and Ma, L. (2003) Direct, dynamic assessment of cell-matrix interactions inside fibrillar collagen lattices. *Cell Motil. Cytoskeleton* **55,** 254–264.
41. Petroll, W. M., Ma, L., and Jester, J. V. (2003) Direct correlation of collagen matrix deformation with focal adhesion dynamics in living corneal fibroblasts. *J. Cell Sci.* **116,** 1481–1491.
42. Vishwanath, M., Ma, L., Jester, J. V., Otey, C. A., and Petroll, W. M. (2003) Modulation of corneal fibroblast contractility within fibrillar collagen matrices. *Invest. Ophthalmol. Vis. Sci.* **44,** 4724–4735.
43. Petroll, W. M., Vishwanath, M., and Ma, L. (2004) Corneal fibroblasts respond rapidly to changes in local mechanical stress. *Invest. Ophthalmol. Vis. Sci.* **45,** 3466–3474.
44. Kaverina, I., Krylyshkina, O., and Small, J. V. (1999) Microtubule targeting of substrate contacts promotes their relaxation and dissociation. *J. Cell Biol.* **146,** 1033–1044.
45. Kaverina, I., Krylyshkina, O., Gimona, M., Beningo, K., Wang, Y. L., and Small, J. V. (2000) Enforced polarisation and locomotion of fibroblasts lacking microtubules. *Curr. Biol.* **10,** 739–742.
46. Rottner, K., Krause, M., Gimona, M., Small, J. V., and Wehland, J. (2001) Zyxin is not colocalized with vasodilator-stimulated phosphoprotein (VASP) at lamellipodial tips and exhibits different dynamics to vinculin, paxillin, and VASP in focal adhesions. *Mol. Biol. Cell* **12,** 3103–3113.
47. Moller-Pedersen, T., Cavanagh, H. D., Petroll, W. M., and Jester, J. V. (1998) Corneal haze development after PRK is regulated by volume of stromal tissue removal. *Cornea* **17,** 627–639.
48. Moller-Pedersen, T., Cavanagh, H. D., Petroll, W. M., and Jester, J. V. (1998) Neutralizing antibody to TGFβ modulates stromal fibrosis but not regression of photoablative effect following PRK. *Curr. Eye Res.* **17,** 736–747.
49. Petroll, W. M., Cavanagh, H. D., Barry-Lane, P., Andrews, P., and Jester, J. V. (1993) Quantitative analysis of stress fiber orientation during corneal wound contraction. *J. Cell Sci.* **104,** 353–363.
50. Edlund, E., Lotano, M. A., and Otey, C. A. (2001) Dynamics of α-actinin in focal adhesions and stress fibers visualized with α-actinin green fluorescent protein. *Cell Motil. Cytoskeleton* **48,** 190–200.

51. Friedl, P., Noble, P. B., and Zanker, K. S. (1995) T lymphocyte locomotion in a three-dimensional collagen matrix. *J. Immunol.* **154,** 4973–4985.
52. Friedl, P., Maaser, K., Klein, C. E., Niggemann, B., Krohne, G., and Zanker, K. S. (1997) Migration of highly aggressive MV3 melanoma cells in 3-dimensional collagen lattices results in local matrix reorganization and shedding of alpha2 and beta1 integrins and CD44. *Cancer Res.* **57,** 2061–2070.
53. Friedl, P., Zanker, K., and Brocker, E.-B. (1998) Cell migration strategies in 3-D extracellular matrix: differences in morphology, cell matrix interactions, and integrin function. *Microsc. Res. Tech.* **43,** 369–378.
54. Hegerfeldt, Y., Tusch, M., Brocker, E.-B., and Friedl, P. (2002) Collective cell movement in primary melanoma explants: plasticity of cell-cell interaction, β1-integrin function, and migration strategies. *Cancer Res.* **62,** 2125–2130.
55. Wolf, K., Mazo, I., Leung, H., et al. (2003) Compensation mechanism in tumor cell migration: mesenchymal-amoeboid transition after blocking of pericellular proteolysis. *J. Cell Biol.* **160,** 267–277.
56. Petroll, W. M., Cavanagh, H. D., and Jester, J. V. (2004) Dynamic three-dimensional visualization of collagen matrix remodeling and cytoskeletal organization in living corneal fibroblasts. *Scanning* **26,** 1–10.

7

Quantitative Analyses of Cell Adhesion Strength

Nathan D. Gallant and Andrés J. García

Summary

Biochemical and mechanical analyses of integrin-mediated cell adhesion have been limited by the inability to apply controlled detachment forces and the inherent complexities of the adhesive process, including cell spreading, integrin clustering, cytoskeletal interactions, and non-uniformly distributed focal complexes. A comprehensive set of techniques to analyze mechanical and biochemical events at the cell–extracellular matrix interface is presented. The spinning disk assay provides a rigorous hydrodynamic assay to measure adhesion strength. Crosslinking/extraction and wet-cleaving biochemical approaches isolate and quantify integrins bound to their ligands and adhesive components recruited to focal adhesions. These techniques provide an experimental framework for the rigorous analysis of cell adhesion.

Key Words: Cell adhesion strength; integrin; focal adhesion; extracellular matrix; vinculin; fibronectin.

1. Introduction

1.1. Cell Adhesion

Cell adhesion to extracellular matrix (ECM) components is necessary for the organization, maintenance, and repair of numerous tissues *(1–3)*. Similarly, cell adhesion to adsorbed proteins or adhesive motifs engineered on surfaces is critical to biomaterials, tissue engineering, and biotechnological applications *(4)*.

Cell adhesion to ECM components, including fibronectin (FN) and laminin, is primarily mediated by the integrin family of heterodimeric receptors *(3)*. Integrin-mediated adhesion is a highly regulated process involving receptor activation and mechanical coupling to extracellular ligands. Bound receptors rapidly associate with the actin cytoskeleton and cluster together to form focal adhesions, discrete supramolecular complexes containing structural (e.g., vinculin, talin, α-actinin) and signaling (e.g., FAK, Src, paxillin) components. Focal

From: *Methods in Molecular Biology, vol. 370:*
Adhesion Protein Protocols, Second Edition
Edited by: A. S. Coutts © Humana Press Inc., Totowa, NJ

adhesions are central elements in the adhesion process, functioning as structural links between the cytoskeleton and the ECM and triggering signaling pathways that direct cell survival, proliferation, and differentiation *(5,6)*.

Although significant progress has been made in understanding the biochemical components of integrin-mediated adhesion, the mechanical aspects of adhesion remain poorly understood because of a lack of robust, quantitative measurement systems, and the inherent complexities of the adhesive process. Nevertheless, mechanical analyses of integrin-mediated cell adhesion to FN have demonstrated a highly regulated, two-stage process involving initial receptor–ligand interactions and subsequent adhesion strengthening and cell spreading *(7–9)*. However, until recently these studies have been limited to short-term adhesion (<60 min) before robust focal adhesions develop. Application of these quantitative approaches to long-term adhesion has been restricted by the inability to apply sufficient forces and the complexity of the strengthening process, including cell spreading, integrin clustering, cytoskeletal interactions, and nonuniformly distributed focal complexes. Recent work from our group establishes a framework for the analysis of adhesion strengthening and the elucidation of the contributions of integrin binding, focal adhesion assembly, and cell spreading *(10,11)*.

1.2. Integrated Analyses of Mechanical and Biochemical Adhesive Events

The methods described in this chapter provide a complementary set of techniques to study adhesion strengthening in terms of mechanical and biochemical events. The spinning disk assay can be used to determine initial as well as long-term adhesion strength *(9–12)*. Because this assay relies on the application of hydrodynamic forces, the cell shape/morphology as well as the size/distribution of adhesive structures influences the forces applied to the cell. In order to eliminate cell-shape-dependent effects, micropatterning techniques have been applied to control cell shape (*see* **Note 1** and **Fig. 1**) *(10,11)*. This chapter does not detail microcontact printing because this technique has been widely used over the last 10 yr and has been reviewed elsewhere *(13,14)*. In addition, a pair of biochemical assays is described that allow the quantification of proteins involved in cell adhesion. The first technique applies a crosslinking/extraction method to measure the number of integrins bound to an ECM-coated substrate. The second method is used to quantify the cytoskeletal proteins that are recruited to basal adhesive structures.

2. Materials

2.1. Cell Adhesion Strength Assay

1. Circular glass cover slips with adherent cells (*see* **Note 2**).

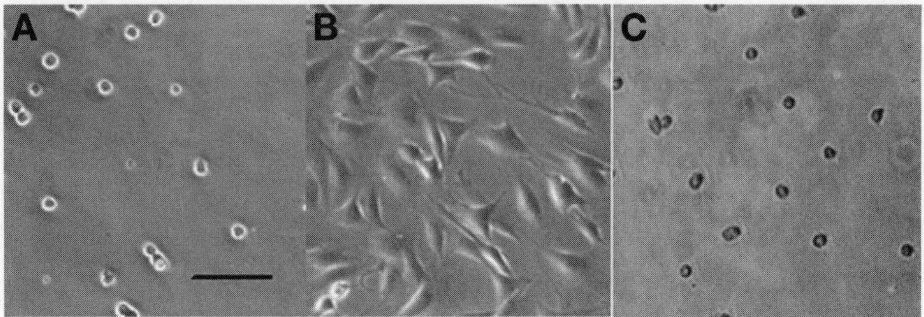

Fig. 1. Phase contrast images of adherent cells on fibronectin at varying times: **(A)** initial adhesion (15 min); **(B)** unpatterned cells spread (16 h); **(C)** micropatterned cells (16 h). Bar, 100 µm.

2. Spinning buffer: 2 m*M* glucose in complete Dulbecco's phosphate-buffered saline solution (DPBS) (*see* **Note 3**).
3. Spinning disk apparatus (*see* **Fig. 2**).
4. 3.7% v/v Formaldehyde in ice-cold DPBS.
5. 1% v/v Triton X-100 in DPBS.
6. Ethidium D homodimer no. 1 (Molecular Probes, Eugene, OR) diluted 1:500 in DPBS.
7. Microscope with an automated stage controller and image analysis software.

2.2. Integrin Binding Assay

1. Substrates with adherent cells.
2. DPBS.
3. Crosslinking solution: 1 m*M* 3,3'-dithiobis(sulfosuccinimidylpropionate) (DTSSP; Pierce Chemical, Rockford, IL) (*see* **Note 4**) and 2 m*M* dextrose in ice-cold DPBS.
4. Quench buffer: 50 m*M* Tris-HCl in DPBS.
5. Extraction buffer: 0.1% sodium dodecyl sulfate (SDS), 350 µg/mL phenylmethyl-sulfonyl fluoride (PMSF), 10 µg/mL leupeptin, and 10 µg/mL aprotinin in DPBS.
6. Reversal buffer: 50 m*M* dithiothreitol (DTT), 0.1% SDS in DPBS (without Ca^{2+} and Mg^{2+}) warmed to 37°C.
7. Microcon centrifugation concentrators.
8. Laemmli sample buffer (2X): 125 m*M* Tris-HCl (pH 6.8), 10% glycerol, 10% SDS, 130 m*M* DTT, 0.01% Bromophenol blue in water.

2.3. Focal Adhesion Quantification Assay

1. Substrates with adherent cells.
2. Wash buffer: DPBS, 2 m*M* PMSF, 1 µg/mL leupeptin, 10 µg/mL aprotinin, 2 m*M* $NaVO_4$.
3. Microcon centrifugation concentrators.

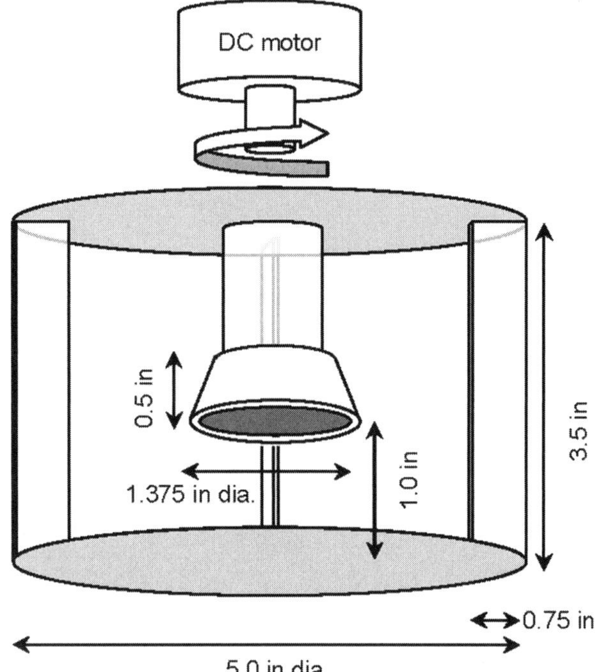

Fig. 2. Schematic of spinning disk device for adhesion strength measurements. Cover slip containing adherent cells is placed on disk head mounted on the end of the motor shaft and submerged in the chamber filled with spinning buffer. Critical dimensions are specified for laminar flow at the sample surface.

4. Laemmli sample buffer (2X): 125 mM Tris-HCl (pH 6.8), 10% glycerol, 10% SDS, 130 mM DTT, 0.01% bromophenol blue in water.
5. Nitrocellulose membrane trimmed slightly larger than the sample area.
6. Glass slides.
7. Small weights as needed.

3. Methods

3.1. Cell Adhesion Strength Assay

Methods for examining cell adhesion strength generally focus on measuring the relative ability of cells to remain attached when exposed to a detachment force. Whereas cell adhesion often refers to simply washing off "non-adherent" cells and counting those that remain, several quantitative adhesion assays have been developed to apply controlled detachment forces to adherent cells. These methods are generally classified according to the type of force applied to detach the cells and can be divided into the categories of (1) micromanipulation, (2) centrifugation, and (3) hydrodynamic force *(15)*. The spinning disk assay, a particular configuration of the hydrodynamic force assay, presents several advantages over other assays, including the ability to generate large detachment forces and the ability to apply a linear range of forces to a large population of cells in a single experiment *(16–18)*. In addition, these shear forces are applied to adherent cells under uniform and constant chemical conditions at the surface.

The spinning disk device consists of a fluid-filled cylinder in which a disk containing the sample substrate is submerged (**Fig. 2**). The sample and disk are connected to a drive shaft protruding into the chamber. The rotation of the disk is driven by a motor, and the speed of the spinning disk is controlled by an optical sensor connected to a tachometer. A system of top and side baffles is used to minimize rotation of the bulk fluid. Dimensions critical to achieve the characterized laminar flow patterns have been highlighted in the schematic. The spinning disk configuration has been extensively studied *(19,20)*, and the flow patterns and mass transport have been characterized electrochemically *(18,21)*. The resulting shear stress (τ) varies linearly with radial distance from zero at the center of the disk and is given by:

$$\tau = 0.800r\,(\rho\mu\omega^3)^{\frac{1}{2}}$$

where r is the radial distance from the center of the disk, ρ and μ are the fluid density and viscosity, and ω is the rotational velocity of the disk.

After spinning, cells that remain adherent are fixed, permeabilized, and stained. Cells are counted at specific radial positions with an automated stage and image analysis software. Sixty-one fields are analyzed per disk and normalized to the cell count at the disk center, for which the applied force is zero. For a cell population with normally distributed adhesion properties, the detachment profile is predicted to fit a sigmoid whose equation is:

$$f = \frac{f_0}{1 + \exp[b(\tau - \tau_{50})]}$$

where f and τ are the experimental adherent cell fraction and disk surface shear stress and f_0, b, and τ_{50} are the fitted zero stress cell fraction, decay slope, and inflection point, respectively. τ_{50} represents the wall shear stress for 50% detachment. This value represents the average detachment shear stress for the population of adherent cells and is used as a measure of adhesion strength. A sample detachment profile is shown in **Fig. 3**.

3.1.1. Procedure for Cell Adhesion Strength Assay

1. Mount one glass cover slip with adherent cells on the vacuum chuck holder at the end of the rotating shaft.

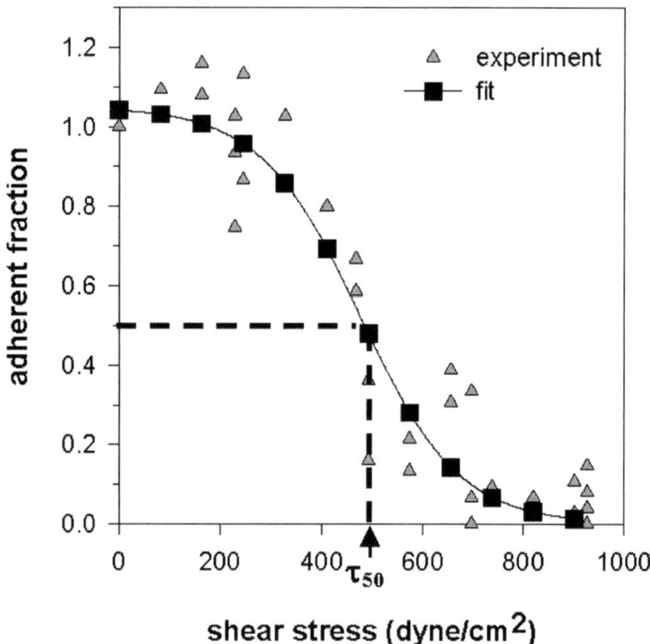

Fig. 3. Cell adhesion strength measurement from a single experiment. The detachment profile shows the fraction of adherent cells as a function of applied shear stress for cells adhering to 5-μm-diameter islands for 16 h. Each data point represents the normalized cell count of one field. Experimental points were fit to a sigmoid to obtain the shear stress for 50% detachment (τ_{50}). (Adapted from **ref. *11*.**)

2. Submerge the disk by inverting into a baffled chamber filled with spinning buffer of defined density and viscosity (*see* **Note 3**).

3. Start the motor and ramp up to the desired speed over 30 s. Spin at a constant speed for 4 min and then decelerate back to zero over 30 s (total spin cycle 5 min). Motor control is achieved with a potentiometer modulating the signal from the direct current power supply. The actual rotational speed of the disk is monitored by an optical sensor.

4. Remove the rotating shaft and spinning disk from the fluid chamber and invert to an upright position.

5. Release the vacuum on the back side of the sample and transfer the disk to a 35-mm dish.

6. Fix cells with 1 mL of cold 3.7% formaldehyde in DPBS for at least 10 min.

7. After all runs are completed, permeabilize cultures for 5 min in 1 mL of 1% Triton X-100.

8. Stain the cell nuclei for counting with ethidium homodimer for 20 min in the dark.

9. Mount the substrates onto microscope slides in preparation for counting.

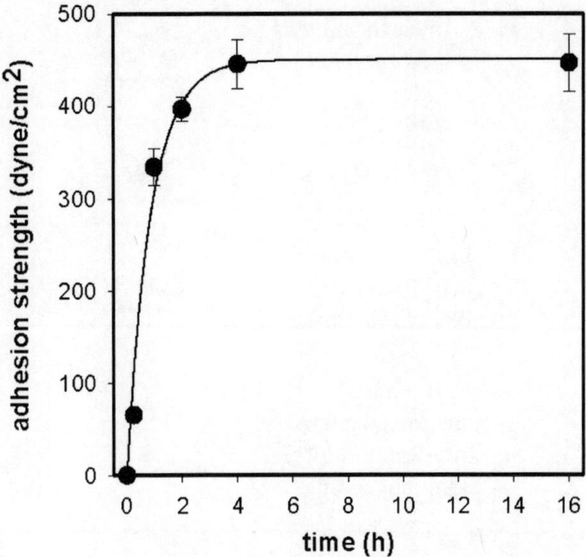

Fig. 4. Evolution of cell adhesion strength over time for cells plated on 5-μm-diameter islands (200 ng/cm² fibronectin). Adhesion strength exhibits rapid increases until reaching steady-state values at 4 h. Exponential curve fit accurately describes data (R^2 > 0.95). (Adapted from **ref. *11*.**)

10. An automated stage controller and image analysis software are used to center the disk and then automatically capture images and count the number of fluorescent cells in 61 fields at known radial positions (*see* **Note 5**).
11. The cell counts are normalized to the count at the center of the disk where zero force was applied, and this adherent fraction remaining after spinning is plotted as a function of the wall shear stress on the disk.
12. These data are then fit to a sigmoid, and the shear stress at which 50% of the cells become detached (τ_{50}) is defined as the mean adhesion strength. Examples of cell detachment profile and evolution of adhesion strength over time are shown in **Figs. 3** and **4**.

3.2. Analysis of Integrin Binding

The number of integrin receptors bound the ECM ligands controls adhesion strength and signaling responses *(11,12,22,23)*. However, measurement of bound receptor numbers is challenging because only a fraction of the total pool of integrin receptors are ligated. This requires the isolation of bound integrins from unligated receptors. The biochemical technique described here provides a simple technique to quantify only those receptors bound to the ECM *(24,25)* (*see* **Fig. 5A**). In this method, bound integrins are crosslinked to their ligand via a chemical reagent with a short spacer arm, thereby excluding any unligated

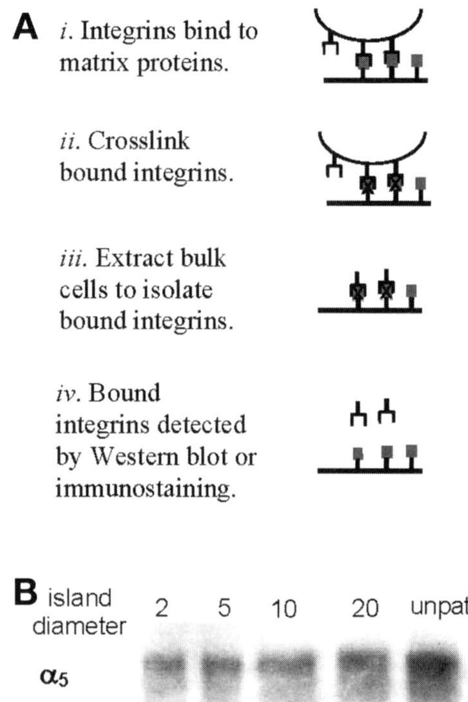

Fig. 5. Quantitative analysis of integrin binding using a **(A)** crosslinking/extraction method consisting of (*i*) integrin binding to adsorbed adhesive protein, (*ii*) crosslinking-bound integrins to the underlying matrix, and (*iii*) detergent extraction to isolate bound integrins for (*iv*) detection by Western blot or immunostaining. **(B)** Western blot of bound α5-integrin subunits for cells on micropatterned islands and unpatterned surfaces. (Adapted from **refs.** *11* and *25*.)

receptors. Detergent extraction of the cell body and unbound receptors leaves behind only those bound and crosslinked integrins that can be subsequently collected by cleaving the crosslinker. The quantity and type of bound integrins is analyzed by Western blot (*see* **Note 6**). Sample results for quantified bound integrins are presented in **Figs. 5B** and **7B**. Control experiments have demonstrated that this approach specifically isolates bound integrins *(24)*.

1. Rinse adherent cells 3X gently with DPBS.
2. Incubate cultures with DTSSP crosslinking solution for 30 min at room temperature (*see* **Note 4**).
3. Quench the remaining crosslinker using a buffer containing Tris-HCl for 5 min.
4. Extract cells with SDS-containing extraction buffer for 5 min (*see* **Note 7**).

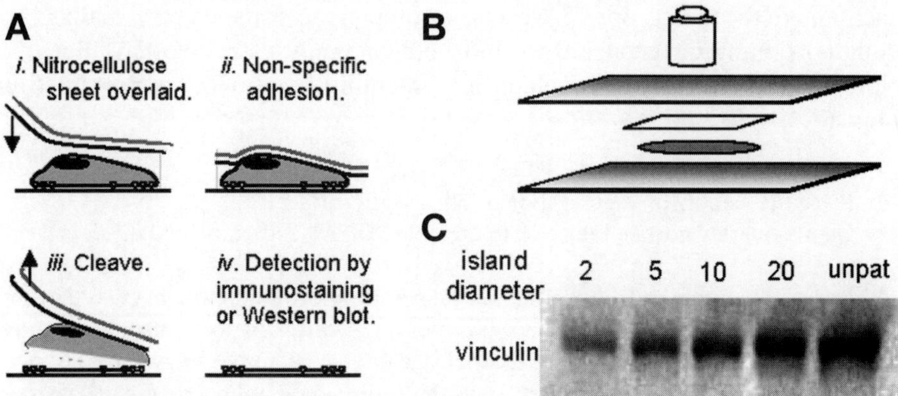

Fig. 6. Quantitative analysis of focal adhesion assembly. (**A**) Schematic of wet-cleaving method: (*i*) overlaying of nitrocellulose onto adherent cells, (*ii*) nonspecific adhesion of membrane-bound proteins to the nitrocellulose, (*iii*) mechanical cleavage of cells to isolate focal adhesion complexes in a plane close to the underlying substrate, and (*iv*) detection by Western blotting or immunostaining. (**B**) Uniform wet cleavage is achieved by sandwiching the sample overlaid with nitrocellulose between two glass slides and adding weight as needed. (**C**) Western blot of vinculin recruited to focal adhesions for cells on micropatterned islands and unpatterned surfaces. (Adapted from refs. *11* and *25*.)

5. Rinse 3X gently with complete DPBS.
6. Reverse the crosslinker by cleaving the disulfide bond of the DTSSP with DTT reversal buffer. Incubate for 30 min at 37°C. Use a minimal volume to limit dilution of proteins with low detectable levels (400 µL is sufficient for 1000 mm^2).
7. It may be necessary to concentrate the sample for detection. Microcon concentration filters can be selected based on the size of the protein being investigated.
8. Concentration of different samples in the same centrifugation cycle will result in all samples having similar concentrations but different volumes. Bring all samples to the same volume with water and sample buffer to 1X to observe differences in protein concentration.
9. Quantify bound integrins by Western blot analysis.

3.3. Analysis of Focal Adhesion Assembly

Quantification of focal adhesion components localized to adhesive plaques requires the isolation of these adhesive structures from bulk cellular components. The wet-cleaving method exposes the cytoplasm of adherent cells in order to quantify focal adhesion assembly (*see* **Fig. 6A**) *(25)*. A nitrocellulose sheet is overlaid onto cells to irreversibly bind protein and membrane components on the apical surface. After a specified time, the nitrocellulose sheet is rapidly removed to mechanically rupture the cell bodies, leaving basal cell structures anchored to the underlying ECM. These components can then be visualized by immunostaining or recovered for biochemical analyses by Western blot (*see* **Note 8**). Sample results for quantified vinculin recruitment are presented in **Figs. 6C** and **7C**.

1. Rinse adherent cells gently with wash buffer.
2. Place the sample on a glass slide or other rigid surface.
3. Gently overlay nitrocellulose on the sample to nonspecifically bind the dorsal cell surface.
4. Sandwich sample and nitrocellulose by covering with a small weight (5–100 g) and a glass slide to distribute the pressure evenly (*see* **Fig. 6B**) for 1 min (*see* **Note 9**).
5. Disassemble the wet-cleavage sandwich and lift the membrane away from the cover slip in a single motion to cleave away the cell fraction not bound to the substrate through adhesive structures.
6. Rinse the substrate gently with wash buffer and aspirate.
7. Solubilize remaining focal adhesions in 1X sample buffer (100 µL is sufficient for 1000 mm^2). Scrape the sample surface to ensure complete solubilization and collect proteins.
8. It may be necessary to concentrate the sample before detection. Microcon concentration filters work well for this purpose and can be selected based on the size of the protein being investigated.
9. Concentration of different samples in the same centrifugation cycle will result in all samples having similar concentrations, but different volumes. Bring all samples to the same volume with water and sample buffer to 1X to observe differences in protein concentration.
10. Quantify focal adhesion components by Western blot analysis.

4. Notes

1. Cell shape is critical to this analysis. Microcontact printing has been used to engineer available adhesive area and control cell spreading, thereby forcing the cells to maintain a spherical shape. This technique is not detailed here, but microcontact printing and other cell patterning techniques have been reviewed elsewhere *(13,14)*.
2. We use 25-mm-diameter glass cover slips in order to obtain a wide range of detachment forces. However, circular shaped substrates are only necessary for the spinning disk adhesion strength assay, whereas any sample shape or area is acceptable for the protein quantification assays. Based on the particular experiment, substrates may be prepared with any protein adsorption or chemical pretreatment and cells adhered prior to adhesion assay.
3. Wall shear stress is dependent on buffer viscosity, and therefore buffer composition may alter applied detachment force. The density and viscosity for the spinning buffer listed is 1.0 g/cm^3 and 1.0 cP.
4. DTSSP must be brought to room temperature before opening the container because condensation will cause rapid hydrolysis. Mix and add solution to sample quickly.

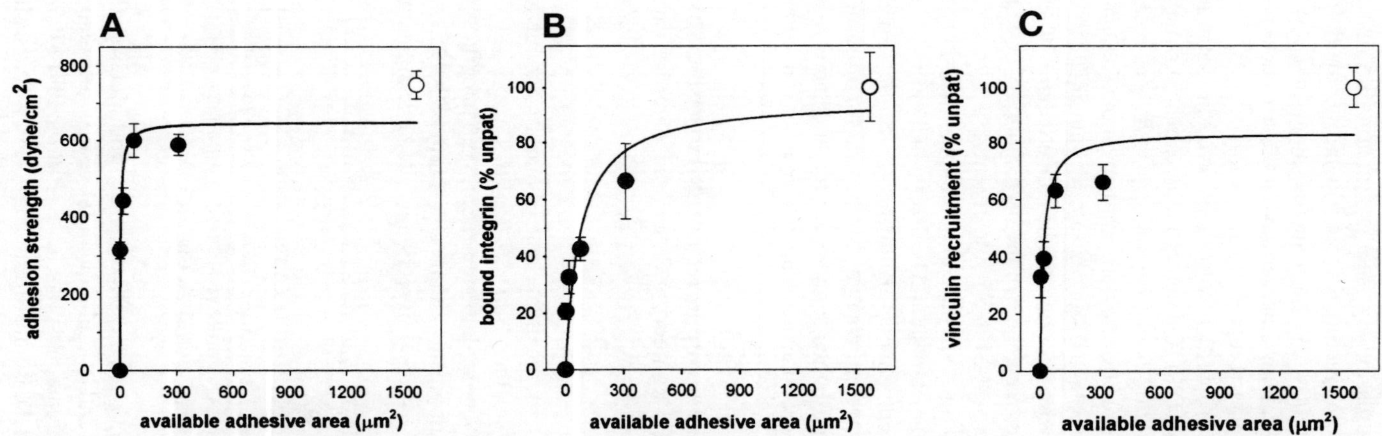

Fig. 7. Relationships between available adhesive area and (**A**) adhesion strength, (**B**) integrin binding, or (**C**) focal adhesion assembly (vinculin recruitment) for micropatterned substrates (closed symbols) and unpatterned (open symbol) at steady state (16 h). Experimental values were fit to hyperbolic curves ($R^2 > 0.90$). (Adapted from **ref. *11***.)

This homobifunctional crosslinking agent (spacer arm 13 Å) reacts with free amine groups. Therefore, only interacting receptor–ligand pairs with free amines will be crosslinked. This reaction works on numerous integrin–ECM bonds, including the $\alpha_5\beta_1$-FN and $\alpha_6\beta_1$-laminin interactions, but a different crosslinking reagent may be necessary for other receptor–ligand pairs.

5. Automation routines were written with Image Pro Plus software (Media Cybernetics, Silver Spring, MD) to first center the disk and then move to 61 specified positions along eight radial paths, capture a picture, and count the fluorescent cells in each image. The cell count data are then output for analysis.
6. Complementary immunostained samples are recommended to confirm observed biochemical trends. Bound integrins can be stained after extraction.
7. The extraction step takes advantage of the high solubility of cell components and near insolubility of intact ECM in SDS detergent.
8. Appropriate weight and time will depend on sample preparation and depth of cleavage desired. These parameters should be determined by immunostaining for cytoskeletal and other cell components. Uniform application of pressure to nitrocellulose membrane and speed of cleaving are critical for uniform cleaving.
9. Complementary immunostained samples are recommended to confirm observed biochemical trends. Focal adhesions can be fixed and stained after cleaving.

Acknowledgments

The authors acknowledge support from NIH (R01-GM065918) and the Georgia Tech/Emory NSF Engineering Research Center for Engineering Living Tissues (EEC-9731643).

References

1. Damsky, C. H. (1999) Extracellular matrix-integrin interactions in osteoblast function and tissue remodeling. *Bone* **25,** 95–96.
2. De Arcangelis, A. and Georges-Labouesse, E. (2000) Integrin and ECM functions: roles in vertebrate development. *Trends Genet.* **16,** 389–395.
3. Hynes, R. O. (2002) Integrins: bidirectional, allosteric signaling machines. *Cell* **110,** 673–687.
4. Garcia, A. J. (2005) Get a grip: integrins in cell-biomaterial interactions. *Biomaterials* **26,** 7525–7529.
5. Sastry, S. K. and Burridge, K. (2000) Focal adhesions: a nexus for intracellular signaling and cytoskeletal dynamics. *Exp. Cell Res.* **261,** 25–36.
6. Geiger, B., Bershadsky, A., Pankov, R., and Yamada, K. M. (2001) Transmembrane extracellular matrix—cytoskeleton crosstalk. *Nat. Rev. Mol. Cell Biol.* **2,** 793–805.
7. Lotz, M. M., Burdsal, C. A., Erickson, H. P., and McClay, D. R. (1989) Cell adhesion to fibronectin and tenascin: quantitative measurements of initial binding and subsequent strengthening response. *J. Cell Biol.* **109,** 1795–1805.
8. Choquet, D., Felsenfeld, D. P., and Sheetz, M. P. (1997) Extracellular matrix rigidity causes strengthening of integrin-cytoskeleton linkages. *Cell* **88,** 39–48.

9. Garcia, A. J., Huber, F., and Boettiger, D. (1998) Force required to break alpha5 beta1 integrin-fibronectin bonds in intact adherent cells is sensitive to integrin activation state. *J. Biol. Chem.* **273,** 10,988–10,993.

10. Gallant, N. D., Capadona, J. R., Frazier, A. B., Collard, D. M., and Garcia, A. J. (2002) Micropatterned surfaces to engineer focal adhesions for analysis of cell adhesion strengthening. *Langmuir* **18,** 5579–5584.

11. Gallant, N. D., Michael, K. E., and Garcia, A. J. (2005) Cell adhesion strengthening: Contributions of adhesive area, integrin binding and focal adhesion assembly. *Mol. Biol. Cell* **16,** 4329–4340.

12. Garcia, A. J., Takagi, J., and Boettiger, D. (1998) Two-stage activation for alpha5 beta1 integrin binding to surface-adsorbed fibronectin. *J. Biol. Chem.* **273,** 34,710–34,715.

13. Kane, R. S., Takayama, S., Ostuni, E., Ingber, D. E., and Whitesides, G. M. (1999) Patterning proteins and cells using soft lithography. *Biomaterials* **20,** 2363–2376.

14. Quist, A. P., Pavlovic, E., and Oscarsson, S. (2005) Recent advances in microcontact printing. *Anal. Bioanal. Chem.* **381,** 591–600.

15. Garcia, A. J. and Gallant, N. D. (2003) Stick and grip: measurement systems and quantitative analyses of integrin-mediated cell adhesion strength. *Cell Biochem. Biophys.* **39,** 61–74.

16. Weiss, L. (1961) The measurement of cell adhesion. *Exp. Cell Res. Suppl.* **8,** 141–153.

17. Horbett, T. A., Waldburger, J. J., Ratner, B. D., and Hoffman, A. S. (1988) Cell adhesion to a series of hydrophilic-hydrophobic copolymers studied with a spinning disc apparatus. *J. Biomed. Mater. Res.* **22,** 383–404.

18. Garcia, A. J., Ducheyne, P., and Boettiger, D. (1997) Quantification of cell adhesion using a spinning disc device and application to surface-reactive materials. *Biomaterials* **18,** 1091–1098.

19. Sparrow, E. M. and Gregg, J. L. (1960) Mass transfer, flow, and heat transfer about a rotating disk. *J. Heat Transfer, Trans. ASME*, 294–302.

20. Levich, V. G. (1962) *Physicochemical Hydrodynamics.* Prentice Hall, Englewood Cliffs, NJ.

21. Serad, G. (1964) Chemical hydrodynamics with a rotating disk. PhD thesis. University of Pennsylvania, Philadelphia, PA.

22. Shi, Q. and Boettiger, D. (2003) A novel mode for integrin-mediated signaling: tethering is required for phosphorylation of FAK Y397. *Mol. Biol. Cell* **14,** 4306–4315.

23. Garcia, A. J. and Boettiger, D. (1999) Integrin-fibronectin interactions at the cell-material interface: initial integrin binding and signaling. *Biomaterials* **20,** 2427–2433.

24. Garcia, A. J., Vega, M. D., and Boettiger, D. (1999) Modulation of cell proliferation and differentiation through substrate-dependent changes in fibronectin conformation. *Mol. Biol. Cell* **10,** 785–798.

25. Keselowsky, B. G. and Garcia, A. J. (2005) Quantitative methods for analysis of integrin binding and focal adhesion formation on biomaterial surfaces. *Biomaterials* **26,** 413–418.

8

Using RNA Interference
to Knock Down the Adhesion Protein TES

Elen Griffith

Summary

RNA interference (RNAi) is a specific and efficient method to knock down protein levels using small interfering RNAs (siRNAs), which target mRNA degradation. RNAi can be used in mammalian cell culture systems to target any protein of interest, and several studies have used this method to knock down adhesion proteins. We used siRNAs to knock down the levels of TES, a focal adhesion protein, in HeLa cells. We demonstrated knockdown of both TES mRNA and TES protein. Although total knockdown of TES was not achieved, the observed reduction in TES protein was sufficient to result in a cellular phenotype of reduced actin stress fibers.

Key Words: TES; RNAi; focal adhesion; actin stress fibers; HeLa.

1. Introduction

TES, originally identified as a candidate tumor suppressor, is an adhesion protein that localizes to focal adhesions, cell–cell contacts, and actin stress fibers *(1–4)*. Overexpression studies have enabled initial characterization of TES, such as identifying subcellular localization and interacting partners *(2–4)*. Some functional data have also been revealed through overexpression studies, for example, TES overexpression results in increased cell spreading and decreased cell motility *(2,4)*. However, to further identify the importance of TES in adhesion complexes, a method to knock down TES function in cells was required, and RNA interference (RNAi) proved to be a suitable method *(5)*.

RNAi is a mechanism that naturally exists in nearly all organisms studied to date, and it can be exploited to knock down a gene in a highly specific and efficient manner. RNAi relies upon small interfering RNA molecules (siRNAs), which mediate degradation of mRNA in a sequence-specific manner. siRNAs,

From: *Methods in Molecular Biology, vol. 370:*
Adhesion Protein Protocols, Second Edition
Edited by: A. S. Coutts © Humana Press Inc., Totowa, NJ

when transfected into a cell, form a complex with various endogenous proteins collectively known as the RNA-induced silencing complex *(6)*. RNA-induced silencing complex recognizes the target mRNA through RNA–RNA interactions and cleaves the mRNA, leading to degradation. Although first discovered in plants and *Caenorhabditis elegans*, more recently RNAi has been utilized to knock down genes in mammalian cells *(7–9)*.

Described here will be the method used for the knockdown of TES in HeLa cells using siRNAs. Also described will be the techniques necessary to assess the knockdown of TES using RNAi, including reverse transcription–polymerase chain reaction (RT-PCR), Western blotting, and immunofluorescence. Although the methods described in this chapter are tailored to the adhesion protein TES, they can be applied to the knockdown of any adhesion protein using RNAi.

2. Materials

2.1. Preparation of siRNAs

1. Single-stranded siRNAs (Proligo, Hamburg, Germany) (*see* **Note 1**).
2. siRNA annealing buffer: 30 mM N-2-hydroxyethylpiperazine-N'-2-ethanesulfonic acid (HEPES) pH 7.5, 100 mM potassium phosphate, and 2 mM magnesium phosphate. Store at room temperature.

2.2. Cell Culture and Transfections

1. Cell culture media: Dulbecco's modified Eagle's media (DMEM) is supplemented with 10% fetal calf serum and 1% L-glutamine (all from Invitrogen, Paisley, UK).
2. OptiMEM-1, trypsin–ethylene diamine tetraacetic acid (EDTA), and phosphate-buffered saline (PBS) (Invitrogen).
3. Serum-free media: DMEM is supplemented with 1% L-glutamine.
4. Transfection reagent: Oligofectamine (Invitrogen).
5. 3X serum media: DMEM is supplemented with 30% fetal calf serum and 1% L-glutamine.

2.3. RT-PCR

1. PBS (Invitrogen).
2. Trizol reagent, chloroform (Sigma, Poole, UK).
3. GeneAMP RNA PCR kit (Perkin Elmer, Buckinghamshire, UK).
4. PCR reagents: 10X PCR buffer, $MgCl_2$, and Taq polymerase (all from Promega, Southhampton, UK).
5. dNTP working solution: prepare 10 mM dNTP stock solution by diluting each of the four dNTPs 1/10 in distilled H_2O (*see* **Note 2**), e.g., 10 μL dATP + 10 μL dCTP + 10 μL dGTP + 10 μL dTTP + 60 μL H_2O (individual 100 mM dNTPs from Promega).
6. PCR primers: TESF; 5'-agcacaaaggatcccaatattcctgc-3'; TESR; 5'-tgccctggatccta agacatcctcttc-3'; GAPDHF; 5'-gtggatattgttgccatcaatgacc-3'; GAPDHR; 5'-ggactc

cacgacgtactcagcgccagc-3' (primers synthesized by Beatson Institute for Cancer Research, Glasgow, UK). Prepare 10 μM working solution for each primer.

7. 1 X TBE (1 L): 10.8 g Tris-base, 5.5 g boric acid, and 4 mL 0.5 M EDTA; make up to 1 L with H_2O.
8. Ethidium bromide from Sigma (St. Louis, MO).

2.4. Western Blotting

1. Cell lysis buffer (10 mL): 50 μL NP40/IGEPAL (Sigma), 500 μL 5 M NaCl, 500 μL 1 M HEPES buffer pH 7.0 (Invitrogen), 100 μL 0.5 M EDTA, 100 μL 100X protease inhibitor cocktail (Calbiochem, Nottingham, UK); make up to 10 mL with H_2O. Aliquot into 1-mL Eppendorf tubes and store at −20°C.
2. Gel apparatus: the following instructions are designed for using the Mini-PROTEAN 3 Cell system (Biorad). However, they can be adapted for other gel running systems.
3. 10% Sodium dodecyl sulfate (SDS) (100 mL): dissolve 10 g SDS in H_2O to a final volume of 100 mL.
4. Acrylamide/bisacrylamide: 30% solution (37.5:1) (Sigma).
5. 10% Ammonium persulfate (1 mL): dissolve 0.1 g in 1 mL H_2O. Prepare fresh each time or store at 4°C for up to 1 wk.
6. Resolving gel (enough for two 0.75-mm gels): mix 4 mL H_2O, 2.5 mL 1.5 M Tris-HCl pH 8.8, 100 μL 10% SDS, and 3.35 mL acrylamide/bisacrylamide solution.
7. Stacking gel (enough for two 0.75-mm gels): mix 3.1 mL H_2O, 1.25 mL 0.5 M Tris-HCl pH 6.8, 50 μL 10% SDS, and 0.65 mL acrylamide/bisacrylamide solution.
8. 4X Protein sample buffer (10 mL): mix 1.05 mL H_2O, 1.25 mL 0.5 M Tris-HCl pH 6.8, 5 mL 50% glycerol, 2 mL 10% SDS, and 0.2 mL 0.5% (w/v) bromophenol blue. Aliquot into 0.95 mL and prior to use add 50 μL β-mercaptoethanol. Store at 4°C for up to 1 mo.
9. Gel running buffer (1 L): 3 g Tris-base, 14.4 g glycine, 1 g SDS; make up to 1 L with H_2O.
10. Rainbow protein markers (Amersham, Buckinghamshire UK).
11. Polyvinylidene difluoride membrane (Millipore, Hertfordshire, UK).
12. Transfer apparatus: the following instructions are designed for using semidry transfer apparatus (Beatson Institute for Cancer Research). However, they can be adapted for other transfer systems.
13. Transfer buffer (1 L): 3 g Tris-base, 3.8 g glycine; make up to 900 mL with H_2O. Finally add 100 mL methanol.
14. TBST (TBS, 0.2% Tween-20) (1 L): 8.8 g sodium chloride, 6.1 g Tris-base; make up to 800 mL with H_2O. Add HCl to pH 7.5 and then make up to 1 L with H_2O. Finally add 2 mL Tween-20 (Sigma).
15. Blocking buffer (TBST, 5% milk) (100 mL): add 5 g skimmed milk powder (Marvel, Spalding, UK) to 100 mL TBST. Store at 4°C for up to 1 wk.
16. Antibodies: mouse anti-TES (Cancer Research UK, Beatson Laboratories), goat anti-actin (Santa Cruz, CA), goat anti-mouse horseradish peroxidase (HRP; Cell

Signaling, Danvers, MA), rabbit anti-goat HRP (Jackson Immunoresearch Laboratories, West Grove, PA).
17. ECL detection reagent (Amersham).
18. Stripping buffer (500 mL): add 7.5 g glycine and 50 mL 10% SDS to 350 mL H_2O. Add HCl to pH 2.5 and then make up to 500 mL with H_2O.

2.5. Immunofluorescence

1. Fixing solution: PBS plus 3.7% formaldehyde.
2. PBS/ammonium chloride: PBS plus 50 mM ammonium chloride.
3. PBS/ Triton X-100: PBS plus 0.5% Triton X-100 (Sigma).
4. Blocking buffer: PBS plus 10% newborn calf serum (Invitrogen).
5. PBS/Tween-80: PBS plus 0.025% Tween-80 (Sigma).
6. Secondary antibody: sheep anti-mouse fluoroisothiocyanate (Jackson Immunoresearch Laboratories).
7. Phalloidin labeled with TRITC (Sigma).
8. Vectastain (Vector Laboratories, Peterborough, UK).

3. Methods

Using RNAi to knock down adhesion proteins has been successfully carried out in a variety of cell lines, including HeLa cells *(8)*. HeLa cells are easy to maintain in cell culture and provide a good starting point for optimizing siRNA transfection conditions. All the methods presented here were optimized to HeLa cells, but in theory any cell line (including nonhuman cells) can be used, although cell lines that are difficult to transfect may be unsuitable as sufficient knockdown may not be achieved.

RNAi is a knockdown rather than knockout method because it is rare to achieve 100% loss of protein from all cells. Because RNAi results in degradation of target mRNA, it is possible to determine the success of an experiment using RT-PCR (*see* **Note 3**). However, it is the loss of functional protein from cells that may reveal a knockdown phenotype, and therefore it is important to also assess protein levels. Protein knockdown can be assessed in two ways, both of which are appropriate when studying adhesion molecules. Western blotting is a method to enable detection of the total level of an individual protein in a population of cells and so can be used to calculate the percentage knockdown of the protein. Immunofluorescence, like Western blotting, uses specific antibodies to detect a protein but, unlike Western blotting, allows individual cells to be visualized. This has the advantage of allowing assessment of the protein knockdown at its specific subcellular localization and therefore enables correlation of the knockdown level and phenotype of the individual cell.

The aim of an RNAi experiment is to knock down protein levels sufficiently to find an observable phenotype in the cells, which may ultimately provide novel insights into protein function. However, it is critically important to include appro-

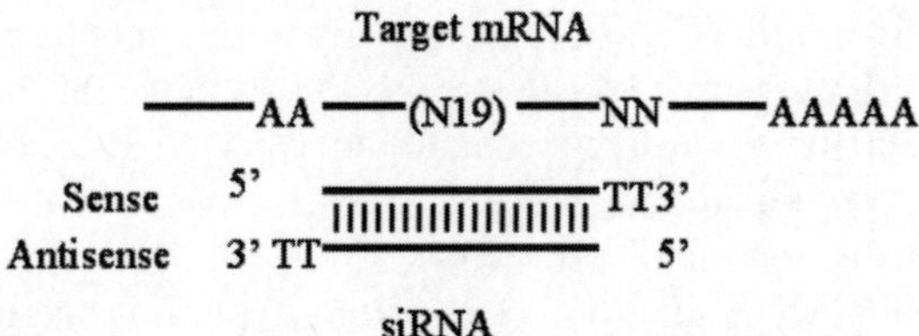

Fig. 1. Design of siRNAs: siRNAs are chosen downstream of an AA dinucleotide with 3' dTdT (TT) overhangs.

priate controls to ensure that the phenotype seen is a real consequence of protein knockdown. Mock transfections containing no siRNA need to be carried out with every experiment. It is also advisable to carry out transfections using siRNA directed against another, unrelated protein to ensure that the phenotype is a result of specific loss of the protein of interest and not simply induction of the RNAi machinery of the cell (*see* **Note 4**). Once knockdown has been achieved and a phenotype is seen, it is advised to use another siRNA to the target gene to ensure that the same phenotype is seen and that off-target effects are not responsible for the phenotype.

3.1. Preparation of siRNAs

1. siRNAs are designed using rules established by Tuschl and coworkers *(8,9)* as follows (*see* **Note 5**). To generate a 21-nucleotide (nt) siRNA, a 19-nt target sequence is chosen immediately downstream of an AA dinucleotide and symmetrical TT 3' overhangs are included (*see* **Fig. 1**) (*see* **Note 6**). For TES RNAi experiments, two siRNAs were designed corresponding to the 19-nt target sequences in positions 91–109 and 777–795 relative to the ATG start codon.
2. Blast search (NCBI database: www.ncbi.nlm.nih.gov/BLAST/) chosen target sequences against EST libraries to ensure that no other genes will be targeted by your siRNAs.
3. Single-stranded siRNAs (*see* **Note 1**) are diluted in annealing buffer to a final concentration of 40 μM. Sense and anti-sense siRNAs are mixed in equal volumes, which gives a final concentration of 20 μM. siRNAs are annealed by incubation at 90°C for 1 min followed by incubation at 37°C for 1 h. Aliquot siRNA duplexes and store at −20°C.

3.2. Cell Culture and Transfections

1. The human cervical carcinoma cell line HeLa is grown in cell culture media at 37°C in 5% CO_2/95% air atmosphere and passaged when approaching confluence with trypsin–EDTA.
2. For RNAi experiments, HeLa cells are seeded into six-well plates to achieve 30–50% confluence at the time of transfection. Usually seeding 1×10^5 cells per well 24 h prior to transfection will result in the correct level of confluence (*see* **Note 7**).

3. Before carrying out transfections, cells are washed: remove media by aspiration; then add 2–3 mL prewarmed serum-free media. Remove this media by aspiration and then add 800 µL serum-free media. Return cells to 37°C incubator while transfection mixes are prepared.

4. A 200-µL transfection mix is prepared:
   a. Mix 3 µL Oligofectamine with 12 µL serum-free media and incubate for 15 min at room temperature (*see* **Note 8**).
   b. Add up to 10 µL each siRNA (*see* **Note 9**) to OptiMEM 1 to give a total 185 µL. Add mix A (15 µL) to mix B (185 µL) and incubate for 15–20 min at room temperature. The transfection mix is then added to the cells, which are then returned to the incubator. After 4 h at 37°C, add 500 µL 3X serum media to the cells to achieve 10% serum in the media.

5. Cells are harvested 24–72 h after transfection (*see* **Note 10**).

3.3. Using RT-PCR to Assess mRNA Knockdown

1. Harvest cells by trypsinization: aspirate media from cells and wash once with 1 mL PBS. Aspirate PBS and add 1 mL trypsin–EDTA to each well. After 5–10 min incubation, gently tap cell culture plate, and once cells have lifted from the bottom of the well, remove cells and add to 2 mL media (with serum) in a 15-mL centrifuge tube. Pellet cells by centrifugation at 1000*g* for 5 min and then aspirate supernatant. Wash cell pellet with 1 mL PBS and repellet cells by centrifugation. Aspirate the supernatant.

2. To islolate RNA add 500 µL of Trizol to each cell pellet and pipet repetitively to lyse cells (*see* **Note 11**). Incubate at room temperature for 5 min. Add 100 µL of chloroform to the sample, shake by hand for 15 s, and then incubate for 2–3 min at room temperature. Centrifuge samples at 12,000*g* for 10 min at 4°C. Remove the upper aqueous phase, transfer to a fresh Eppendorf tube, and add 250 µL isopropanol. Incubate samples for 10 min at room temperature and then centrifuge at 12,000*g* for 10 min at 4°C. Remove the supernatant and then wash the RNA pellet by adding 500 µL of 75% ethanol and vortexing, and then centrifuge at 7500*g* for 5 min at 4°C. Remove the supernatant and then air-dry the RNA pellet for 10 min at room temperature. Resuspend the RNA pellet in 20 µL of RNase-free water and incubate for 10 min at 60°C. The integrity and concentration of the RNA can be determined by agarose gel electrophoresis (*see* **step 5**). Store RNA samples at –80°C.

3. Carry out reverse transcription to make cDNA using the GeneAmp RNA PCR kit, which contains all the necessary reagents. Thaw RNA samples on ice and denature 3 µL of each RNA sample by incubation for 10 min at 65°C. While RNA is denaturing, set up 20-µL reverse transcription reactions containing the following: 4 µL MgCl$_2$ (25 m*M*), 2 µL 10X PCR buffer, 2 µL dATP (10 m*M*), 2 µL dCTP (10 m*M*), 2 µL dGTP (10 m*M*), 2 µL dTTP (10 m*M*), 1 µL RNase inhibitor, 1 µL MuLV reverse transcriptase, 1 µL random hexamers primer (*see* **Note 12**). Finally, add 3 µL of denatured RNA sample and mix. Reverse transcription is carried out by incubating samples for 10 min at room temperature, then 42°C for 1 h, 99°C for 5

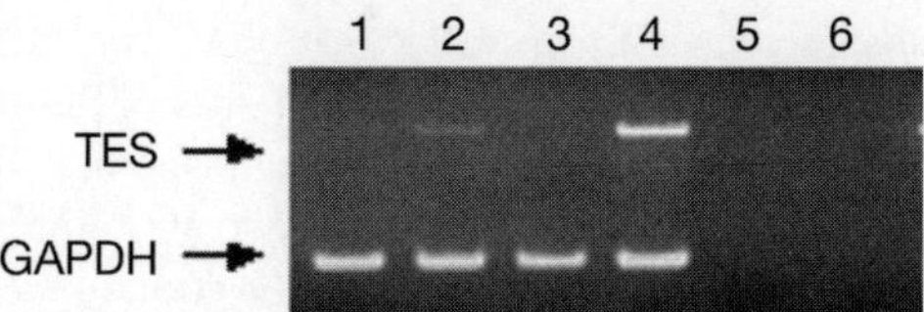

Fig. 2. TES knockdown measured by reverse transcription–polymerase chain reaction (RT-PCR). RT-PCR was carried out to assess TES mRNA levels in HeLa cells that were treated with each TES siRNA individually (lanes 1 and 2), both TES siRNAs (lane 3), or mock treated (lane 4). Lanes 5 and 6 are controls for RT (no RNA) and PCR (no cDNA), respectively.

min, and 5°C for 5 min. Store cDNA samples at –20°C. A control reaction containing no RNA is carried out.

4. PCR reactions are carried out using cDNA generated from the reverse transcription as the template. The 50-µL PCR reactions are set up so that each reaction contains the following: 5 µL 10X PCR buffer, 4 µL MgCl$_2$ (25 mM), 1 µL dNTP working solution (10 mM), 2 µL each primer (10 µM) (TESF, TESR), 0.2 µL control primers (10 µM) (GAPDHF, GAPDHR), 2 units Taq polymerase, 1 µL cDNA (from **step 3**); this is made up to a final volume of 50 µL with distilled H$_2$O. When multiple PCRs are set up, a master mix is prepared containing all the above reagents except the cDNA, which is added individually to each PCR tube. Usually along with experimental samples, a control containing no cDNA is included. The PCR reactions are carried out as follows: 94°C for 30 s, 55°C for 30 s, 72°C for 1 min; this is repeated for 30 cycles (*see* **Note 13**).

5. PCR reactions are then run out on a 1% agarose gel in 1X TBE buffer at 100 V for 45 min. Stain the gel for 10 min in TBE containing ethidium bromide at a concentration of 0.5 µg/mL and visualize using UV light. An example of TES RT-PCR is given in **Fig. 2.**

3.4. Using Western Blotting to Assess Protein Knockdown

1. Harvest cells as in **Subheading 3.3., step 1**.
2. Prepare whole cell extract as follows: lyse cells by adding 100 µL of cell lysis buffer to cell pellet and pipet repetitively. Incubate on ice for 30 min and then centrifuge at 12,000*g* for 10 min at 4°C. Remove supernatant (whole cell extract) and transfer to fresh tube, which should be kept on ice; store samples at –20°C.
3. The following method is for the preparation of two 10% 0.75-mm gels using 10-well combs: take 10 mL resolving gel mix and just prior to pouring add 50 µL ammonium persulfate solution and 5 µL TEMED. Pour the resolving gel, leaving enough room for stacking gel on top. Overlay the resolving gel with 100% ethanol. Once the resolving gel is set (this will take about 30 min), prepare the stacking gel as follows: take 5 mL stacking gel mix and just prior to pouring add 25 µL ammonium persulfate solution and 5 µL TEMED. Pour off the ethanol and then

pour the stacking gel mix on top of the resolving gel, remembering to add the combs. The stacking gel should polymerize in about 30 min.

4. Prepare protein samples as follows: prepare equal amounts of total protein for each sample (usually 10–50 µg is sufficient) and mix with 4X protein sample buffer. Boil the samples for 5 min at 95–100°C and then cool on ice. Briefly spin down in a microcentrifuge to collect samples at bottom of tube.

5. Set up gel tank by adding gel running buffer to the upper reservoir and to the tank (about 400 mL in total). Load protein samples (up to 20 µL) into the wells, and load 5 µL protein markers in one well. Run the gel for approx 45 min at 200 V.

6. Dissassemble gel and place gel in transfer buffer for 10 min prior to transfer. Set up semidry transfer as follows. Prepare polyvinylidene difluoride membrane prior to setup: soak membrane in methanol for 2 min and then rinse with H_2O. Place membrane in transfer buffer. Wet each piece of Whatmann paper in transfer buffer before assembly. Place three pieces of Whatmann paper onto lower plate of transfer apparatus, place preprepared membrane on top of Whatmann paper, then place gel on top of membrane. Finally, place three more pieces of Whatmann paper on top of gel. Ensure that no bubbles of air are trapped between the layers by rolling over gently with a falcon tube. Place lid onto transfer apparatus and transfer for approx 30 min at 20 V.

7. After disassembling the transfer, the membrane should be blocked in blocking solution for approx 1 h at room temperature with gentle rocking.

8. The primary antibody, e.g., TES 1C10, is diluted 1/100 in 5–10 mL blocking solution, added to membrane, and incubated overnight at 4°C with gentle rocking (*see* **Note 14**).

9. The primary antibody is removed and the membrane washed once briefly in TBST. The secondary antibody (anti-mouse HRP) is then diluted 1/3000 in 5–10 mL blocking solution, added to the membrane, and incubated for 1 h at room temperature. The membrane is washed three times for 10 min at room temperature in TBST.

10. Chemiluminescent detection is carried out using the ECL reagent: remove membrane from TBST and dry briefly on tissue. Place membrane onto cling film on a flat surface. Mix equal volumes of solutions 1 and 2 and then immediately pour over membrane, ensuring that the surface area of the membrane is entirely covered. Leave for 1 min and then drain by holding edge of membrane against cling film. Wrap membrane up in cling film and place in cassette ready for exposure to film. Expose membrane to X-ray film for between 10 s and 30 min to achieve desired level of exposure.

11. Reprobe membrane with actin antibody as a loading control (*see* **Note 15**). First strip membrane by incubating in stripping buffer for up to 1 h at room temperature with gentle rocking (*see* **Note 16**). Wash membrane once briefly in TBST.

12. Repeat **steps 8–11** using actin primary antibody diluted 1/5000 and using anti-goat HRP secondary antibody diluted 1/3000. An example of a TES and actin Western blot is shown in **Fig. 3**.

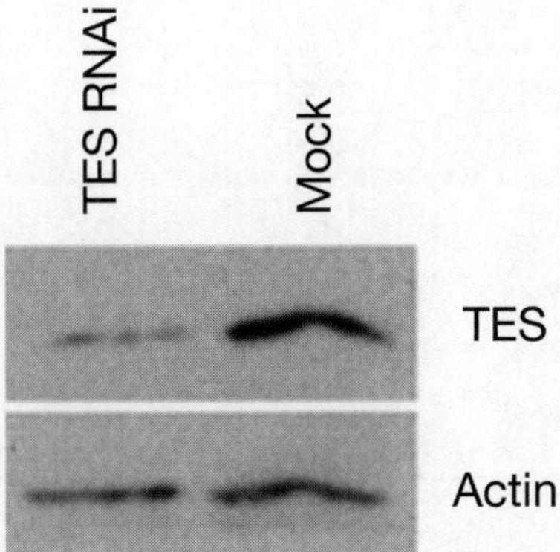

Fig. 3. TES knockdown measured by Western blot. Western analysis was performed to assess TES protein levels in HeLa cell extracts from TES siRNA and mock-treated cells. Blots were probed with an anti-TES antibody (1C10) **(upper)** and reprobed with an anti-actin antibody as a loading control **(lower)**. (Reproduced with the permission of the publisher, John Wiley and Sons, Wiley-Liss, and **ref. 5**.)

3.5. Using Immunofluorescence to Assess Protein Knockdown

1. Set up and transfect cells as in **Subheading 3.2.**, but when setting up cells place two sterile 16-mm glass cover slips into each well of the six-well plate and ensure they are flat on the bottom of the well.
2. Forty-eight hours after transfection, remove media and wash cover slips once with PBS. Fix cells onto cover slips by adding approx 2 mL fixing solution to each well and incubate for 15 min at room temperature. Wash cover slips three times for 5 min in PBS (*see* **Note 17**).
3. Incubate cover slips for 15 min in PBS/ammonium chloride solution and then wash twice in PBS.
4. Incubate cover slips for 5 min in PBS/Triton X-100 solution.
5. Block cover slips by incubation in blocking buffer for 30 min.
6. Incubate cover slips in primary antibody, e.g., TES 1C10 diluted 1/100 in blocking buffer at 4°C overnight (*see* **Note 14**). Antibody incubations can be carried out in a large volume such as 2 mL, or, alternatively, to minimize antibody use, 80 µL can be placed onto the cover slip. If using only 80 µL, then place your plate into a humid box for the incubation period and take care not to disturb it.
7. Wash cover slips four times 4 min in PBS plus 0.025% Tween-80.
8. Incubate cover slips in secondary antibody, e.g., sheep anti-mouse fluoroisothiocyanate diluted 1/200 in blocking buffer for 30 min at room temperature.
9. To stain the actin cytoskeleton, incubate cover slips in phalloidin-TRITC diluted 1/500 in blocking buffer for 30 min at room temperature in the dark.
10. Wash cover slips four times for 4 min in PBS plus 0.025% Tween-80.

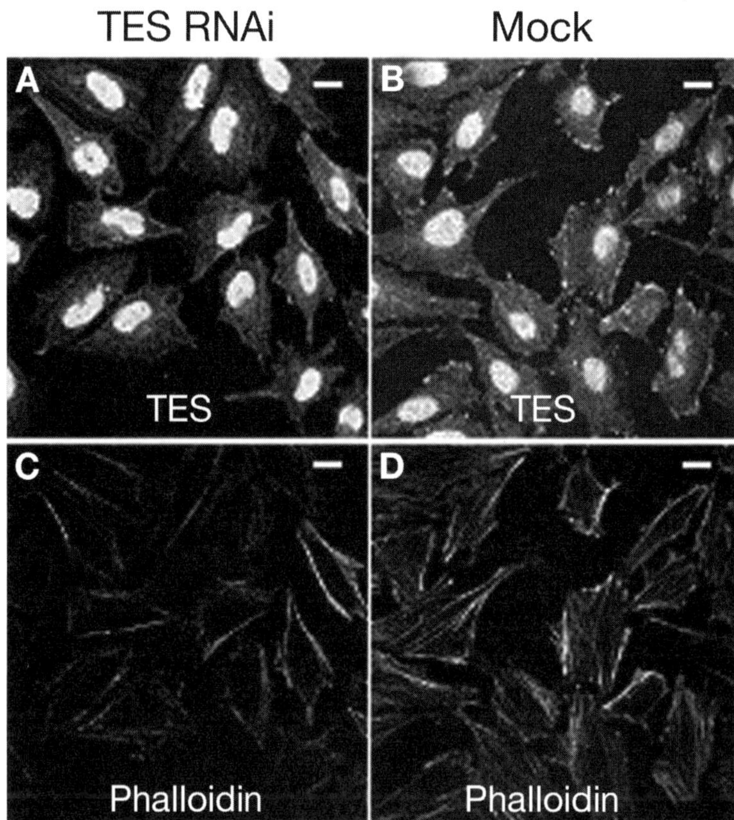

Fig. 4. TES knockdown and the resulting cellular phenotype were assessed using immunofluorescence. HeLa cells were treated with TES siRNAs **(A,C)** or mock RNAi **(B,D)**. Forty-eight hours posttransfection cells were fixed and stained with an anti-TES antibody (1C10) and a fluoroisothiocyanate-labeled mouse secondary **(A,B)**. Cells were co-stained with TRITC-labeled phalloidin to visualize the actin stress fibers **(C,D)**. Actin staining revealed that TES knockdown results in reduced actin stress fibers. Bar = 10 µm. (Reproduced with the permission of the publisher, John Wiley and Sons, Wiley-Liss, and **ref. 5**.)

11. Mount cover slips by inverting onto a drop of Vectastain on a glass slide, remove excess liquid, and seal using nail polish. Store cover slips at 4°C in the dark.
12. Cells can be visualized using a confocal or fluorescence microcope at ×63 magnification. Immunofluorescence images of TES knockdown and the resulting cellular phenotype are shown in **Fig. 4**.

4. Notes

1. Many companies now sell a range of custom-made and predesigned siRNAs, including Proligo, Dharmacon, Ambion, and Qiagen. Various options are available

    when ordering siRNAs, and it is usually possible to order your siRNA duplexes preannealed.

2. Use distilled H_2O for all steps in this chapter that require the use of H_2O.

3. Northern blotting is a more quantitative method that can be used to assess mRNA knockdown as an alternative to RT-PCR. If available, quantitative PCR may also be used.

4. Many researchers choose a scrambled version of their siRNA as a control, but this may not induce the RNAi machinery and so might not fully rule out the possibility that the phenotype is a result of nonspecific RNAi effects. Now many companies that sell siRNAs also sell preevaluated siRNAs to use as experimental controls. Also, there are many validated targets in the literature that can be used for this control.

5. Although the Tuschl siRNA design rules have been followed by many researchers, there now exist many design tools to help decide which target sequence to choose. Often these design tools are available on the web sites of companies that sell siRNAs.

6. The 2-nt 3' overhangs can be synthesized using UU or dTdT, but using dTdT may be cheaper. Other overhangs can also be used; for advice consult the siRNA user guide at www.rockefeller.edu/labheads/tuschl/sirna.

7. HeLa cells are maintained in cell culture without the use of antibiotics. However, if antibiotics are usually used in cell culture, then use media without antibiotics when setting up RNAi experiments.

8. There are now many different transfection reagents specifically for transfecting siRNAs; these can be purchased from the major companies who manufacture transfection reagents and siRNAs. If using a different transfection reagent to Oligofectamine, follow manufacturer's guidelines for transfection protocols.

9. It is necessary to determine the least amount of siRNA that is effective to achieve the sufficient protein knockdown. Usually 10–100 n*M* final concentration of siRNA is sufficient. Using multiple siRNAs may reduce the individual concentration of each siRNA needed.

10. When optimizing siRNA transfections, a variety of time points can be measured to determine at what point the maximum knockdown is seen. Most researchers find that at 24–72 h posttransfection a sufficient knockdown is seen, although longer time points can also be used.

11. Use Trizol reagent with caution and avoid any contact with the skin. It is advised to wear protective gloves and eye protection.

12. Two primers are included in the kit: random hexamers and Oligo d(T)16. Either can be used for reverse transcription.

13. The numbers of cycles used can be reduced to increase the sensitivity of the RT-PCR for measuring mRNA knockdown.

14. Alternatively, primary antibodies can be incubated for 1 h at room temperature.

15. If your protein of interest is similar in size to actin (43 kDa), it may be easier to use an alternative protein as a loading control.

16. Stripping times may vary for different antibodies. Stripping can be checked by reincubation with the secondary antibody followed by ECL detection. Alternatively,

if your protein of interest is significantly different in size from actin, you can cut and remove the section of the membrane containing your protein of interest and then probe the remaining section for actin.

17. At this stage cover slips can be stored in PBS at 4°C for several weeks, and immune staining can be carried out later.

References

1. Tobias, E. S., Hurlstone, A. F. L., MacKenzie, E., and Black, D. M. (2001) The TES gene at 7q31.1 is methylated in tumors and encodes a novel growth-suppressing LIM domain protein. *Oncogene* **20,** 2844–2853.
2. Coutts, A. S., MacKenzie, E., Griffith, E., and Black, D. M. (2003) TES is a novel focal adhesion protein with a role in cell spreading. *J. Cell Sci.* **116,** 897–906.
3. Garvalov, B. K., Higgins, T. E., Sutherland, J. D., et al. (2003) The conformational state of Tes regulates its zyxin-dependent recruitment to focal adhesions. *J. Cell Biol.* **161,** 33–39.
4. Griffith, E., Coutts, A. S., and Black, D. M. (2004) Characterisation of chicken TES and its role in cell spreading and motility. *Cell Motil. Cytoskeleton* **57,** 133–142.
5. Griffith, E., Coutts, A. S., and Black, D. M. (2005) RNAi knockdown of the focal adhesion protein TES reveals its role in actin stress fiber organisation. *Cell Motil. Cytoskeleton* **60,** 140–152.
6. Meister, G. and Tuschl, T. (2004) Mechanisms of gene silencing by double-stranded RNA. *Nature* **431,** 343–349.
7. Huppi, K., Martin, S. E., and Caplen, N. J. (2005) Defining and assaying RNAi in mammalian cells. *Mol. Cell* **17,** 1–10.
8. Harborth, J., Elbashir, S. M., Bechert, K., Tuschl, T., and Weber, K. (2001) Identification of essential genes in cultured mammalian cells using small interfering RNAs. *J. Cell Sci.* **114,** 4557–4565.
9. Elbashir, S. M., Harborth, J., Lendeckel, W., Yalcin, A., Weber, K., and Tuschl, T. (2001) Duplexes of 21-nucleotide RNAs mediate RNA interference in cultured mammalian cells. *Nature* **411,** 494–498.

9

Assaying Calpain Activity

Neil O. Carragher

Summary

The calpains represent a well-conserved family of intracellular proteases that exhibit broad substrate specificity and consequently influence many cellular processes. Calpain activity regulates the turnover of integrin-linked focal adhesions, which controls cell adhesion to extracellular matrix substrates, cell migration across such substrates, as well as the signalling output from the associated integrin receptors. Thus, calpain activity regulates both the physical interaction and biochemical communication between cells and the extracellular environment that is vital for normal cell function. Modulation of calpain activity is associated with a number of human diseases, and recent studies have identified calpain activity as an important therapeutic target in several disease areas. These studies have driven the development of in vitro and live-cell-based assays for high-throughput identification and evaluation of calpain inhibitors. However, many of the unique features of calpain activity devalue the physiological relevance of existing assays. In this chapter we describe a modified approach that monitors calpain activity by taking into account several of the factors that control calpain activity and substrate specificity under physiological conditions. The development of such second-generation assays may help to identify more effective calpain intervention strategies.

Key Words: Calpain; Src; focal adhesion kinase; talin; cleavage; sodium dodecyl sulfate–polyacrylamide gel electrophoresis; phosphorylation; fluorescence resonance energy transfer; biosensor.

1. Introduction

The calpain family of intracellular calcium-dependent cysteine proteases are composed of several ubiquitously expressed and tissue-specific isoforms (*1*). In contrast to other proteases that result in complete protein digestion, calpain activity in most cases results in the limited proteolytic cleavage of protein substrates into discernable fragments that often retain biological activity. Therefore, calpain activity not only controls the total levels of protein substrates but

From: *Methods in Molecular Biology, vol. 370:*
Adhesion Protein Protocols, Second Edition
Edited by: A. S. Coutts © Humana Press Inc., Totowa, NJ

can also modify their intracellular location, signaling function, and scaffolding capabilities. The calpains exhibit broad substrate specificity and as a consequence are implicated in regulating a number of physiological processes, including cell-cycle progression, transcription, apoptosis, and cell migration. Recent evidence indicates that calpain-mediated cleavage of the focal adhesion component talin is the critical rate-limiting step that controls the turnover of integrin-linked focal adhesions *(2)*. Integrins are a family of heterodimeric transmembrane receptors that function as components of focal adhesion complexes that represent the main sites of attachment and biochemical communication between cells and their surrounding extracellular matrix substrates *(3)*. Integrin receptors are therefore vital for normal cell homeostasis, which permits cell survival, proliferation, and cell migration. Calpains were initially implicated in focal adhesion regulation by the observation that the calpain 2 isoform localizes to discrete focal adhesion sites in several cell types *(4)*. Several components of the focal adhesion complex, including focal adhesion kinase (FAK), paxillin, talin, α-actinin, and β_3-integrin, were subsequently identified as calpain substrates *(5–8)*. Further studies utilizing pharmacological inhibitors and calpain-null cells demonstrate a critical role for calpain in focal adhesion turnover required for optimal cell migration *(9–11)*.

A unique feature of the calpains is that they do not cleave proteins at a defined consensus sequence. Instead, it has been speculated that the three-dimensional conformation of protein substrates determines calpain recognition and defines the site of cleavage. Thus, posttranslational modifications that alter protein conformation, such as phosphorylation, can both positively and negatively regulate substrate susceptibility to protein cleavage by calpain *(12–14)*. Several synthetic peptide substrates of calpain have been generated that allow quantification of calpain activity. These reagents can be applied to cells, cell extracts, or in vitro experiments. While these assays provide a relatively sensitive measurement of calpain activity, their effectiveness is circumvented by significant limitations. First, peptide substrates often lack specificity for calpain proteolytic activity. Second, these assays measure the cleavage of a synthetic peptide and not physiological substrates, which may be dependent on other factors such as posttranslational modification. Third, existing peptide substrates only measure the total amount of calpain activity in cells and do not quantify the localized activation of calpain that has been reported to take place at discrete subcellular locations such as the plasma membrane *(15)*. The methods described here for assaying calpain activity are designed to specifically evaluate the impact that posttranslational modification (in this example phosphorylation) has on the susceptibility of focal adhesion substrates to undergo calpain-mediated cleavage.

This chapter also outlines a fluorescence resonance energy transfer (FRET)-based assay designed to monitor localized calpain activity in live cells. FRET describes the transfer of energy from a donor fluorophore to an appropriately positioned acceptor fluorophore over a limited distance, resulting in subsequent fluorescence emission by the acceptor fluorophore. A critical requirement for FRET is that the distance between donor and acceptor fluorophores be within the range of 1–10 nm. This requirement has resulted in FRET being widely used as an application to monitor protein–protein interactions. Recent studies have employed fusion proteins possessing donor and acceptor fluorogenic protein pairs suitable for FRET on either side of a proteolytic cleavage site *(16,17)*. FRET generated by such proteins is lost in response to their proteolytic cleavage, and thus when transfected into cells, these fusion proteins represent useful biosensors of proteolytic activity. Such biosensors of calpain activity, when in live cells, are also subject to various modes of posttranslational modification that may influence their sensitivity to proteolysis by calpain.

2. Materials

2.1. In Vitro Calpain Assay

1. Anti-[354-534]FAK mouse monoclonal antibody (BD Pharmingen, Franklin Lakes, NJ).
2. Anti-mouse immunoglobulin conjugated to agarose beads (Sigma, St. Louis, MO).
3. Purified calpain 1 (porcine erythrocytes) or calpain 2 (porcine kidney) (Calbiochem: Merck Biosciences, Darmstadt, Germany).
4. Human recombinant Src (Upstate Biotechnology, Inc., Lake Placid, NY).
5. Modified RIPA lysis buffer: 1% Triton-X 100, 0.5% NP40, 150 mM NaCl, 10 mM Tris-HCl (pH 7.4), 1 mM ethylene glycol tetraacetic acid, 10 mM Na pyrophosphate, 100 µM Na orthovanadate, 1 mM phenylmethylsulfonyl fluoride, 10 µg/mL leupeptin, and 10 µg/mL aprotinin.
6. Src kinase buffer: 100 mM Tris-HCl (pH 7.2), 5 mM MgCl$_2$, 2.5 mM MnCl$_2$, 1 mM dithiothreitol, 500 µM ATP.
7. Calpain assay wash buffer: 50 mM Tris-HCl (pH 7.4), 10 mM CaCl, 30 mM NaCl.
8. Calpain reaction buffer: 50 mM Tris-HCl (pH 7.4), 10 mM CaCl, 30 mM NaCl, 5 mM β-mercaptoethanol.
9. 4X Sodium dodecyl sulfate (SDS) sample buffer: glycerol 40% (v/v), 250 mM Tris-HCl, pH 6.8, SDS 8% (w/v), bromophenol blue 0.4% (w/v), β-mercaptoethanol 8% (v/v).

2.2. SDS-Polyacrylamide Gel Electrophoresis

1. Modular Mini-PROTEAN II electrophoresis equipment (Bio-Rad, Hercules, CA).
2. *Bis*-acrylamide 30% w/v (37.5:1) (Anachem, Bedfordshire, UK).
3. Separating gel: *bis*-acrylamide (5.62 mL), distilled water (9.08 mL), 1.5 M Tris-HCl, pH 8.8 (10 mL), 10% (w/v) SDS (200 µL), TEMED (20 µL), 10% ammonium persulfate (100 µL).

4. Stacking gel: *bis*-acrylamide (2.6 mL), distilled water (12.2 mL), 0.5 *M* Tris-HCl, pH 6.8 (5 mL), 10% (w/v) SDS (200 μL), TEMED (20 μL), 10% ammonium persulfate (100 μL).
5. Running buffer (1X): Tris-base (7.5 g), glycine (36 g), SDS (2.5 g). Make up to 2 L with distilled water; adjust to pH 8.3.
6. Transfer buffer (1X): glycine (14.4 g), Tris-base (3.03 g), methanol (100 mL); make up to 1 L with distilled water.
7. Nitrocellulose membrane (Hybond-ECL, Amersham-Pharmacia, Piscataway, NJ).
8. Prestained molecular-weight rainbow markers (Amersham-Pharmacia).
9. Phosphate-buffered saline (PBS).
10. PBS–Tween containing 0.5% nonfat dry milk: dissolve 0.1% (v/v) polyoxyethylenesorbitan monolaurate (Tween-20, Sigma) and 0.5% nonfat dry milk (w/v) (Marvel, Premier Beverages) in PBS.
11. Blocking buffer: PBS–Tween containing 5% (w/v) nonfat dry milk.
12. Primary rabbit polyclonal anti-$^{2\text{-}17}$N-FAK antibody (Santa Cruz Biotechnology Inc., Santa Cruz, CA).
13. Secondary anti-rabbit horseradish peroxidase-conjugated antibody (New England Biolabs Inc., Ipswich, MA).

2.3. FRET Biosensor Monitoring of Calpain Activity in Live Cells

1. A mammalian cDNA expression vector encoding a fusion protein composed of two green fluorescent protein variants that are suitable as a FRET donor and acceptor pair, spanned by an amino acid sequence composed of a known calpain cleavage site (*see* **Subheading 3.3.**).
2. Transfection reagent (Superfect, Qiagen).
3. Fluorescent microscope fitted with charge-coupled device camera or confocal microscope imaging system.
4. Image analysis software (Slidebook: Intelligent Imaging Systems).

3. Methods

This chapter describes modified in vitro and cellular assays that are suitable for measuring calpain activity while simultaneously taking into account posttranslational modifications that may influence susceptibility of protein substrates to calpain-mediated cleavage. Although no consensus cleavage site appears to direct proteolysis mediated by calpain, calpain does appear to cleave each of its protein substrates preferentially at a specific site. These same protein substrates may also be further processed by calpain cleavage at other, less optimal sites. Thus, proteolytic cleavage fragments initially generated by calpain activity may be subject to further proteolysis by calpain in response to longer incubation times or higher concentrations of calpain enzyme. Therefore, for each calpain substrate being studied in an in vitro calpain cleavage assay, it is important that both the concentration of calpain and reaction time be optimized in order to distinguish the optimal (initial) cleavage events from the subsequent

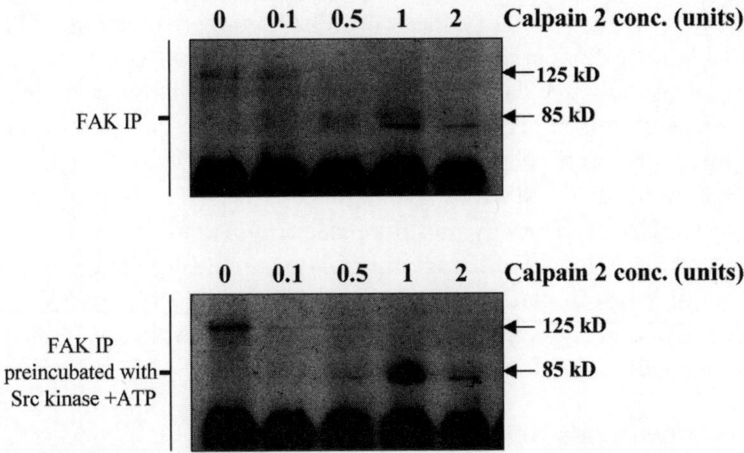

Fig. 1. Concentration-dependent cleavage of focal adhesion kinase (FAK) by calpain in vitro. Immunoprecipitated FAK, untreated (**upper panel**) or phosphorylated by Src (**lower panel**), was incubated with increasing concentrations of purified calpain 2 (0.1–2 U per sample, 0 = no calpain control). Intact native FAK protein and calpain-generated FAK fragments were detected by sodium dodecyl sulfate–polyacrylamide gel electrophoresis followed by immunoblotting with anti-N-terminal FAK antibody. Native FAK protein at 125 kDa and the calpain-generated 85 kDa N-terminal proteolytic fragment of FAK can clearly be observed. 0.5 activity units of calpain 2 are required to significantly diminish native levels of nonphosphorylated FAK (lane 3, upper panel). In contrast, only 0.1 activity units of calpain 2 is necessary to significantly reduce native levels of Src-phosphorylated FAK (lane 2, lower panel). These data suggest that while nonphosphorylated FAK is cleaved efficiently by calpain 2, FAK, which is phosphorylated by Src, is significantly more susceptible to calpain-mediated proteolysis. The levels of the 85-kDa cleavage fragment generated by calpain decrease following incubation with the highest concentrations of calpain 2 (2 U) indicate that the 85-kDa N-terminal cleavage fragment of FAK is further processed by calpain 2.

suboptimal cleavage. Posttranslational modification of protein substrates such as phosphorylation has previously been demonstrated to modify susceptibility to calpain-mediated proteolysis *(12–14)*. In the assay described below the influence that phosphorylation of FAK by the nonreceptor tyrosine kinase Src has on susceptibility of FAK to calpain cleavage is examined (**Fig. 1**). However, a similar approach can be employed to examine the effect of other phosphorylation events or, indeed, other posttranslational modifications have on calpain-mediated cleavage of other substrates (*see* **Note 1**). In addition, while the end-product protein fragments derived by calpain proteolysis are generally stable, they may only be present in low abundance as a result of limited cleavage of the native protein substrate. Therefore, every effort to increase the sensitivity of methods employed to identify these protein fragments (such as SDS-PAGE, described below) must be taken into consideration (*see* **Note 3**).

Many of the cell-free calpain activity assays or calpain activity assays performed on extracted cell lysates may not be appropriate or sensitive enough to detect physiological regulation of calpain activity that is restricted to specific subcellular regions. Thus, imaging calpain activity in live cells using a combination of modern microscopy and fluorescence imaging techniques may provide more physiologically relevant information on the regulation and targeting of calpain activity. Regarding this, we also describe FRET-based assays that have recently been employed to detect the activity of proteases, including calpain, in live cells.

3.1. In Vitro Calpain Activity Assay

1. Immunopreciptation of endogenous FAK protein from 1 mg of cell lysate extracted from proliferating mouse embryo fibroblasts: wash cells twice with PBS and lyse in modified RIPA lysis buffer. Lyse cells on ice for 30 mins, and then clarify by centrifugation (15,000g) at 4°C. Remove clarified lysate supernatants to a clean Eppendorf tube and incubate overnight with 2.5 µg of anti-FAK monoclonal antibody ([354-534]FAK, BD Pharmingen) at 4°C with constant mixing. Add 30 µL of anti-mouse immunoglobulin conjugated to agarose beads (Sigma) to each sample and incubate for a further 2 h at 4°C with constant mixing.
2. For samples subjected to phosphorylation by Src prior to incubation with calpain, wash serially (3X) with lysis buffer followed by washes (2X) with Src kinase buffer. Following the second wash with Src kinase buffer, resuspend samples in 50 µL of Src kinase buffer and add recombinant Src (0.25 U/µL). Incubate samples for 60 min at 30°C. Subsequently incubate the reaction mixture at 56°C for 15 min to inactivate Src. Samples not subject to Src-dependent phosphorylation were treated as above minus the addition of Src.
3. Serially wash samples with calpain assay wash buffer (2X) and calpain reaction buffer (2X). Following the final wash in calpain reaction buffer, resuspend samples in 150 µL of calpain reaction buffer. Aliquot samples into 5X 30-µL individual samples. Add to each sample increasing concentrations of purified calpain 2 enzyme ranging from 0 (control) to 2 U. Incubate calpain reactions at 30°C for 30 min.
4. Terminate each calpain reaction by the addition of 7.5 µL of 4X SDS sample buffer. Subsequently boil samples at 100°C for 10 min and store at −20°C prior to analysis by SDS-PAGE.

3.2. SDS-PAGE

1. This protocol of SDS-PAGE is based on the method first described by Laemmli, who originally used this technique to study the cleavage of structural proteins *(18)*. The protocol described is designed to examine cleavage fragments of FAK that are generated by calpain, using the mini PROTEAN II apparatus from BioRad

(**Fig. 1**). Altering the concentration of acrylamide in the gels may be required to visualize calpain-generated fragments of other substrates (*see* **Note 2**).

2. Prepare the separating gel solution as described in **Heading 2.** to give a 7.5% (w/v) acrylamide gel. Apply approx 4 mL of the separating gel solution to the glass plate sandwich assembled onto the mini PROTEAN II gel apparatus.

3. Overlay top of separating gel with an even layer of 50% (v/v) methanol, 200 µL per gel. Allow gel to polymerize for 45 min.

4. Wash off 50% methanol layer using copious amounts of distilled water. Remove excess water by blotting with 3MM filter paper.

5. Prepare stacking gel solution as described in **Heading 2.** and apply on top of separating gel. Insert a 1-mm 15-tooth Teflon comb. Allow stacking gel solution to polymerize for 45 min.

6. Carefully remove Teflon comb from the gel. Remove glass plate sandwich from casting rack, and attach to the Bio-Rad mini-gel running apparatus. Fill middle chamber with running buffer and partially fill outer chamber with running buffer.

7. Thaw molecular-weight markers and FAK immunoprecipitation samples from the calpain activity assay on ice and centrifuge briefly prior to loading 15 µL of sample and 10 µL of molecular weight standards into individual wells of the gel.

8. Run gels at 150 V (constant voltage) until the 30-kDa "orange" marker has run off the gel.

9. Following gel electrophoresis, remove gels and immunoblot separated protein and proteolytic fragments onto a nitrocellulose filter using the Bio-Rad mini gel "wet blot" transfer apparatus.

10. Following transfer, place nitrocellulose membranes into a dish with PBS–Tween containing 0.5% nonfat dry milk (w/v) and wash for 5 min on an orbital shaker at room temperature.

11. Block nonspecific antibody binding to the membrane by incubating the membrane in blocking buffer for 30 min at room temperature with constant agitation.

12. Dilute a specific antibody recognizing the N-terminal portion of FAK (Santa Cruz Biotechnology Inc.) in PBS–Tween-containing 0.05% (w/v) nonfat dry milk to obtain an ideal working concentration (1:1000 dilution). Incubate the primary antibody solution with membrane overnight at 4°C with constant agitation.

13. Following primary antibody incubation, wash membranes four times in PBS–Tween at room temperature on shaker.

14. Incubate membrane with suitable primary antibody-relevant (anti-rabbit) horseradish peroxidase-conjugated secondary antibody, diluted 1:5000 in PBS–Tween-containing 0.05% (w/v) nonfat dry milk. Secondary antibody was incubated with membranes for 1 h on shaker at room temperature.

15. Following secondary antibody incubation, wash membranes four times in PBS–Tween-containing 0.5% nonfat dry milk.

16. Analyze antibody binding using the enhanced chemiluminescent (ECL) detection system and reagents (Amersham-Pharmacia) according to manufacturer's instructions.

17. Briefly blot membrane to remove excess ECL reagents and expose to autoradiographic hyperfilm ECL (Amersham-Pharmacia) in autoradiographic hyperfilm cas-

settes (Amersham-Pharmacia) for variable exposure times prior to automatic development of film.

3.3. FRET Analysis

1. Using standard molecular biology techniques, construct a mammalian expression vector encoding the in-frame fusion of enhanced yellow fluorescence protein, then an amino acid sequence consisting of a known calpain cleavage site, followed by enhanced cyan fluorescence protein. Alternative fluorescent proteins may also be used (*see* **Note 4**). A suitable amino acid sequence for calpain 1 has previously been described and consists of 10 amino acids identical to the calpain-sensitive site in the 11th spectrin repeat of α-spectrin, flanked by glycine and serine residues (GSGSGQ QEVYGMMPRDGSG) *(16)*. An amino acid sequence suitable for calpain 2-mediated cleavage would be the recently described cleavage site on talin *(2)* consisting of (STVLQQQYNR) (*see* **Fig. 2**).
2. In order to confirm that the expressed FRET fusion protein is a suitable biosensor for calpain activity, the purified fusion protein should be subjected to the in vitro calpain activity assay followed by SDS-PAGE, as described above. This analysis may be used to confirm that the fusion protein is cleaved by calpain in addition to indicating whether it may be specific for a particular calpain isoform. The substitution of other proteases for calpain in the above assay conditions will further confirm whether the fusion protein is a specific biosensor for proteolytic activity mediated by calpain rather than other proteases.
3. A fusion protein FRET biosensor for calpain activity containing the α-II spectrin cleavage site for calpain 1 has been successfully transfected into Cos-7 and N2A cells using the superfect transfection reagent (Qiagen) *(16)*. The same fusion protein has also been transfected into cultured embryonic hippocampal neurons using calcium phosphate precipitation (Promega) *(16)*.
4. Following transfection, cells may be incubated with a variety of agonists known to promote calpain activity such as epidermal growth factor or a calcium ionophore and/or cell-permeable calpain inhibitors to confirm whether the fusion protein is effectively monitoring calpain activity.
5. Images detecting FRET can be acquired using either a fluorescent microscope attached to a charge-coupled device camera or a microscope attached to a confocal imaging system.
6. Image analysis of FRET involves three basic operations: (1) subtraction of background autofluorescence, (2) quantitation of fluorescence intensity, and (3) calculation of acceptor/donor fluorescence emission ratios (FRET ratios). A variety of software packages can be used for this analysis, including Slidebook, which has been demonstrated to be suitable for monitoring calpain-mediated changes in FRET *(16)*.

4. Notes

1. In the in vitro calpain activity assay described in this chapter, endogenous FAK immunopreciptated from mouse embryo fibroblasts was used as the calpain substrate. Other calpain substrates immunoprecipitated from other cell lines are also

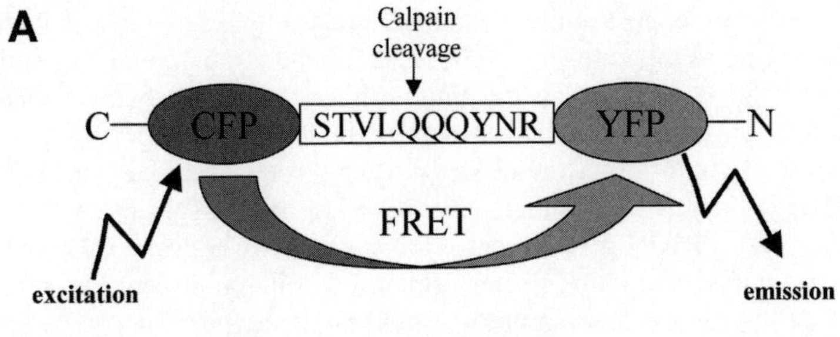

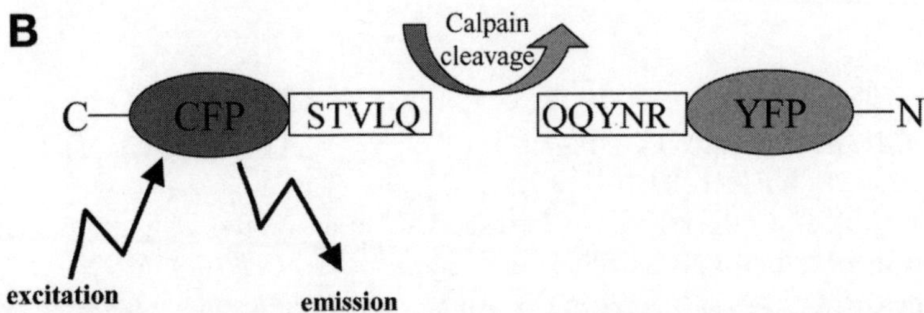

Fig. 2. Design of a fluorescence resonance energy transfer (FRET)-based biosensor to monitor calpain activity in live cells. Calpain-mediated cleavage of talin has recently been demonstrated to be a critical rate-limiting step in focal adhesion turnover *(2)*. The design of a FRET-based biosensor of calpain activity consisting of cyan fluorescence protein (CFP) and yellow fluorescence protein (YFP) fluoroproteins separated by the talin cleavage site for calpain (STVLQQQYNR) represents a useful development for monitoring calpain activity in live cells. (**A**) While intact, following excitation of CFP, the calpain biosensor will emit fluorescence from both CFP and, as a consequence of FRET, also YFP. (**B**) Following calpain-mediated cleavage of the talin sequence, FRET will no longer occur and only CFP will emit fluorescence. Such changes in fluorescence emission can be monitored in cells to study calpain activity. FRET biosensors may also be tagged with membrane or focal adhesion targeting domains to monitor calpain activity at specific subcellular locations.

suitable for analysis. Alternatively, purified or recombinant proteins may also be suitable substrates if the native protein remains intact. However, when interpreting such in vitro data, it should be taken into account that alterations in tertiary protein structure generated by purification or inherent to recombinant protein may alter both susceptibility to calpain as well as the cleavage pattern.

2. The distribution and size of proteolytic fragments generated by calpain vary widely depending on each specific calpain substrate. Therefore, to ensure that all fragments are visualized by SDS-PAGE, samples may have to be run on alternate gels with distinct concentrations of acrylamide. For example, to observe small fragments generated by calpain, it may be necessary to run samples on "high-percentage" acrylamide gels or gradient gels. To observe larger fragments it may be necessary to run samples for a longer period on "low-percentage" acrylamide gels.

3. Quite often, long exposure times may be required to observe low abundant proteolytic fragments separated by SDS-PAGE. The addition of nonfat dry milk powder into buffers used to wash nitrocellulose filters and antibody dilution buffers is intended to limit the appearance of nonspecific bands that appear after such long exposures. The addition of nonfat dry milk in these buffers may not be required when using other antibodies that detect different protein substrates.

4. The design of a FRET-based biosensor for calpain activity described above utilizes the green fluorescent protein variants enhanced cyan fluorescence protein and enhanced yellow fluorescence protein as donor and acceptor fluorophores, respectively. However, other suitable donor and acceptor FRET pairs may also be used.

References

1. Goll, D. E., Thompson, V. F., Li, H., Wei, W., and Cong, J. (2003) The calpain system. *Physiol Rev.* **83,** 731–801.
2. Franco, S. J., Rodgers, M. A., Perrin, B. J., et al. (2004) Calpain-mediated proteolysis of talin regulates adhesion dynamics. *Nat. Cell Biol.* **6,** 977–983.
3. Hynes, R. O. (1992) Integrins: versatility, modulation, and signaling in cell adhesion. *Cell* **69,** 11–25.
4. Beckerle, M. C., Burridge, K., DeMartino, G. N., and Croall, D. E. (1987) Colocalization of calcium-dependent protease II and one of its substrates at sites of cell adhesion. *Cell* **51,** 569–577.
5. Cooray, P., Yuan, Y., Schoenwaelder, S. M., Mitchell, C. A., Salem, H. H., and Jackson, S. P. (1996) Focal adhesion kinase (pp125FAK) cleavage and regulation by calpain. *Biochem. J.* **318,** 41–47.
6. Carragher, N. O., Levkau, B., Ross, R., and Raines, E. W. (1999) Degraded collagen fragments promote rapid disassembly of smooth muscle focal adhesions that correlates with cleavage of pp125(FAK), paxillin, and talin. *J. Cell Biol.* **7,** 619–630.
7. Selliah, N., Brooks, W. H., and Roszman, T. L. (1996) Proteolytic cleavage of alpha-actinin by calpain in T cells stimulated with anti-CD3 monoclonal antibody. *J. Immunol.* **156,** 3215–3221.
8. Pfaff, M., Du, X., and Ginsberg, M. H. (1999) Calpain cleavage of integrin beta cytoplasmic domains. *FEBS Lett.* **460,** 17–22.
9. Huttenlocher, A., Palecek, S. P., Lu, Q., et al. (1997) Regulation of cell migration by the calcium-dependent protease calpain. *J. Biol. Chem.* **272,** 32,719–32,722.
10. Carragher, N. O., Fincham, V. J., Riley, D., and Frame, M. C. (2001) Cleavage of focal adhesion kinase by different proteases during SRC- regulated transformation and apoptosis: Distinct roles for calpain and caspases. *J. Biol. Chem.* **276,** 4270–4275.
11. Dourdin, N., Bhatt, A. K., Dutt, P., et al. (2001) Reduced cell migration and disruption of the actin cytoskeleton in calpain-deficient embryonic fibroblasts. *J. Biol. Chem.* **276,** 48,382–48,388.

12. Bi, R., Rong, Y., Bernard, A., Khrestchatisky, M., and Baudry, M. (2000) Src-mediated tyrosine phosphorylation of NR2 subunits of NMDA receptors protects from calpain-mediated truncation of their C-terminal domains. *J. Biol. Chem.* **275,** 26,477–26,483.

13. Huang, C., Tandon, N. N., Greco, N. J., Ni, Y., Wang, T., and Zhan, X. (1997) Proteolysis of platelet cortactin by calpain. *J. Biol. Chem.* **272,** 19,248–19,252.

14. Nicolas, G., Fournier, C. M., Galand, C., et al. (2002) Tyrosine phosphorylation regulates alpha II spectrin cleavage by calpain. *Mol. Cell. Biol.* **22,** 3527–3536.

15. Glading, A., Uberall, F., Keyse, S. M., Lauffenburger, D. A., and Wells, A. (2001) Membrane proximal ERK signaling is required for M-calpain activation downstream of epidermal growth factor receptor signaling. *J. Biol. Chem.* **276,** 23,341–23,348.

16. Vanderklish, P. W., Krushel, L. A., Holst, B. H., Gally, J. A., Crossin, K. L., and Endelman, G. M. (2000) Marking synaptic activity in dendritic spines with a calpain substrate exhibiting fluorescence resonance energy transfer. *Proc. Natl. Acad. Sci.* **97,** 2253–2258.

17. Mahajan, N. P., Harrison-Shostak, D. C., Michaux, J., and Herman, B. (1999) Novel mutant green fluorescent protein protease substrates reveal the activation of specific caspases during apoptosis. *Chem. Biol.* **6,** 401–409.

18. LaemmLi, U. K. (1970) Cleavage of structural proteins during assembly of the head of bacteriophage T4. *Nature* **227,** 680–685.

10

Analysis of Focal Adhesions and Cytoskeleton by Custom Microarray

Matthew J. Dalby and Stephen J. Yarwood

Summary

Focal adhesions and the cell cytoskeleton (intermediate filaments, microfilaments, microtubules) are involved in mechanotransduction—both direct (transduction of mechanical forces to the nucleus) and indirect (transduction of chemical signaling cascades to the nucleus). Thus, observation of changes in focal adhesion and cytoskeletal organization can be invaluable in research such as drug treatments and medical material testing in vitro.

Here we describe how to stain human fibroblasts for vinculin (located to focal adhesions), actin (microfilaments), tubulin (microtubules), and vimentin (intermediate filaments) and how to perform custom microarray experiments. Comparative analysis of the immunofluorescence and array data should allow the researcher to build up a global picture of changes to both direct and indirect mechanotransduction through the cytoskeleton from focal adhesions.

Key Words: Focal adhesions; focal contacts; cytoskeleton; intermediate filaments; microfilaments; microtubules; mechanotransduction; custom microarray.

1. Introduction

Cells may respond to environmental cues through changes in their focal adhesions and rearrangement of their cytoskeleton. Such cues may be chemical, e.g., different extracellular matrix proteins or drug treatment, or topographical and chemical, e.g., introduction of implants into the body. cDNA microarrays allow us an excellent opportunity to "fish" for important information in a wide variety of cell responses. The use of custom arrays (or glass arrays) is widespread and provide good value for the money. However, caution must always

From: *Methods in Molecular Biology, vol. 370:*
Adhesion Protein Protocols, Second Edition
Edited by: A. S. Coutts © Humana Press Inc., Totowa, NJ

be applied when using microarrays, and backup methodologies should also be applied. Traditionally, quantitative polymerase chain reaction (PCR) is used.

In our work we have used arrays from the Ontario Microarray Centre to study fibroblast response to biomaterials *(1,2)*. Therefore, we will concentrate on detailed description of fibroblast culture and hybridization of mRNA to the Ontario slides. It should be noted that, given the pitfalls we will mention, these methods can easily be adapted for other cell types and custom arrays.

When considering artificial cues delivered by biomaterials (as is our experience), changes in adhesion and cytoskeletal morphology go hand in hand with changes in gene expression data. We therefore provide a morphological approach to examine cell responses.

Changes in adhesion morphology are the starting point for changes in many adhesion pathways, specifically mechanotransductive pathways *(3)*. Mechanotransduction is generally considered to have two forms: direct and indirect. Direct mechanotransduction considers changes in cytoskeleton (driven through mechanical changes in adhesion size/localization) transducing changes in nucleus morphology *(4)*.

It is believed that the cytoskeleton can form interconnected networks forming a mechanically integral network from adhesion to the nucleus involving microfilaments and microtubules localized to focal adhesions and intermediate filaments connecting directly to the lamin intermediate filaments of the nucleoskeleton *(5,6)*. It is further believed that interphase chromosomes may have relative consistency of position within the nucleus *(7–9)* and that changes in lamin nucleoskeleton may alter chromosome positioning, resulting in changes in the probability of gene transcription *(10)*. Microscopy of the cyto- and nuclear skeletons is very useful for looking at these processes.

Indirect mechanotransduction is incredibly complex, consisting of signaling cascades (with adhesions as the start point). Such cascades may involve kinase signaling (e.g., focal adhesion kinase, myosin light chain kinase), and many others), G-protein signaling (e.g., the Ras family of Rho, Rac, and cdc42 involved in cell spreading, sensing, and motility), and ion signaling (e.g., calcium) *(11–14)*. To complicate matters further, these transduction pathways are intimately tied to changes in cytoskeletal (notably microfilament) arrangement. Thus, it can be seen that the use of cDNA microarrays, which measure many genes simultaneously, is useful in the dissection of these pathways. Also, the combination of microscopy and microarray represents a rapid and powerful toolbox for observation of these pathways.

Two recommended reference texts on the background science of focal adhesion morphology and signaling are Bershadsky et al. *(15)* and Burridge and Chrzanowska-Wodnicka *(3)*.

2. Materials

2.1. Cell Culture

1. Human fibroblasts (h-tert BJ1, Clontech).
2. Complete medium: 71% Dulbecco's modified Eagle's medium (Sigma, UK), 17.5% medium 199 (Sigma, UK), 9% fetal calf serum (Life Technologies, UK), 1.6% 200 mM L-glutamine (Life Technologies, UK), and 0.9% 100 mM sodium pyruvate (Life Technologies, UK). For a 500-mL bottle of Dulbecco's modified Eagle's medium, discard 100 mL and replace with 100 mL of medium 199. Discard 50 mL and add 50 mL of fetal calf serum. Next add 10 mL of 200 mM L-glutamate and 5 mL of 100 mM sodium pyruvate. This should be done under sterile conditions (*see* **Note 1**).
3. Trypsin: 0.7 mL of 2.5% trypsin (Gibco Life Technologies) into 20 mL of Versine.
4. Versine: 8 g NaCl, 0.4 g KCl, 1 g D-glucose, 2.38 g N-2-hydroxyethylpiperazine-N'-2-ethanesulfonic acid (HEPES), 0.2 g ethylene diamine tetraacetic acid (EDTA), 2 mL 0.5% phenol red in 1 L water. Adjust to pH 7.5, aliquot, and autoclave (all from Sigma Aldrich).
5. HEPES saline: 8 g NaCl, 0.4 g KCl, 1 g D-glucose, 2.38 g HEPES, 2 mL 0.5% phenol red in 1 L water. Adjust to pH 7.5, aliquot, and autoclave (all from Sigma Aldrich).

2.2. Immunohistochemistry

1. Monoclonal, anti-human (raised in mouse), primary antibodies to vinculin (hvin1), tubulin (tub 2.1), vimentin (V9) (all from Sigma Aldrich). Store at –70°C (*see* **Note 2**).
2. Biotinylated secondary antibody to mouse (BA 2000, Vector Laboratories).
3. Fluoroisothiocyanate (FITC) conjugated streptavidin (BA 5001, Vector Laboratories).
4. Vectorshield with DAPI fluorescent mountant (SA 5001, Vector Laboratories).
5. Phalloidin-Rhodamine fluorescent probe for actin (Molecular Probes). Reconstitute, as directed, in methanol and store at –20°C.
6. Phosphate-buffered saline (PBS): dissolve one tablet in 200 mL distilled water.
7. 4% Formaldehyde: dilute 10 mL (of 38% formaldehyde) in PBS. Add 2 g sucrose (Sigma Aldrich).
8. 0.5% Tween-20: dilute 0.5 mL of Tween-20 in 99.5 mL of PBS.
9. 1% Bovine serum albumin (BSA): dissolve 1 g of BSA in 100 mL of PBS.
10. Permeablizing buffer: dissolve 10.3 g sucrose, 0.292 g NaCl, 0.06 g MgCl$_2$ (hexahydrate), 0.476 g HEPES in 100 mL PBS. Adjust pH to 7.2 and then add 0.5 mL Triton X (all from Sigma Aldrich).
11. Unless specified, store at 4°C (*see* **Note 3**).
12. Standard microscope slides and 22 × 50 mm and 13 mm diameter cover slips (Fisher).

2.3. Microarray

1. 1718 gene arrays (1.7 k human arrays) from Ontario Microarray Centre (http://www.microarrays.ca/).

2. RNA extraction kit from Qiagen (Rneasy [no. 7104]). You will need ethanol and mercaptoethanol (Sigma Aldrich) in order to activate some of the reagents in the kit.
3. Reverse transcription: 5X first strand buffer, mRNA primer (12–18 mers), 20 m*M* dATP, dTTP, and dGTP, 2 m*M* dCTP, Cy3, and Cy5 (Amersham), 0.1 *M* dithiothreitol, Superscript II reverse transcriptase. Unless stated, these are from Gibco Life Technologies.
4. Stop reaction: 50 m*M* EDTA (pH 8.0) (Sigma Aldrich).
5. 10 *N* NaOH (Sigma Aldrich).
6. 5 *M* Acetic acid (Sigma Aldrich).
7. Isopropanol (Sigma Aldrich).
8. Hybridization: EasyHyb (Roche); 10 mg/mL yeast tRNA; 10 mg/mL salmon sperm DNA. Unless stated, these are from Gibco Life Technologies.
9. Washing: 1X saline sodium citrate (SSC), 0.1% sodium dodecyl sulfate (Sigma Aldrich).
10. 22 × 50 mm cover slips (Fisher).

3. Methods

Prior to starting, there are a number of considerations, including

1. Material on which to culture.
2. Cell seeding density.
3. Time of culture period.

It is important that you be able to compare and contrast results from the arrays with results from cytoskeletal observation. However, there are difficulties associated with this. In order to get the best images of the cytoskeleton, it is best that the cells be separated from one another, but to get the best microarray results you need as much RNA as possible. This tends to come from cells at 80–90% confluency, i.e., just before they slow down proliferation. This is because there is normally no amplification step used in microarray as there is in PCR. Amplification can be performed, but it adds to the cost and may hide some more subtle changes.

In our work we have cultured the cells to get the optimal images and the optimal array results. We are forced to do this because we work with biomaterials and this can restrict our sample size; if we could get very large samples, we could grow the cells to the same level of confluency, i.e., one suitable for microscopy, and be able to extract sufficient mRNA for microarray at the same culture time.

These considerations are discussed throughout this section, and you will need to think of a suitable protocol for your cells and your treatment.

3.1. Cell Culture

1. Culture the fibroblasts (or cell of choice, noting that all the systems described here are for human cells and would have to be adapted for other species) in 75-cm^2 cell

culture flasks in 10 mL of complete medium. The cells should be maintained in a 37°C environment with 5% CO_2. If the cells are being cultured for more than a few days, the medium should be replaced twice a week under sterile conditions (e.g., class II safety cabinet).

2. For microscopy, the cells should be trypsinized from the flask and placed on the material of interest or a cover slip for chemical treatment (we suggest 13 mm diameter). For this, remove the media from the flask, wash with 10 mL HEPES/saline, and add 2–3 mL of the trypsin/versine solution for 5 min. Once the 5 min is over, knock the side of the flask to fully dislodge the cells. Pipet out the trypsin/cell mixture and make up to 10 mL in a centrifuge tube with complete media. Centrifuge at 672*g* for 4 min to produce a cell pellet. This can be resuspended (by passing through a fine-gage needle) in 10 mL of media. The cells can be counted for seeding requirements with a hemocytometer at this point.

3. For microarray, larger growth areas will be needed (*see* **Note 8**). Thus, you may want to grow on larger cover slips/areas of test material.

4. Place the materials/cover slips (sterile—at least ethanol washed and rinsed with sterile HEPES/saline) into a tissue culture plate (e.g., a 24-well plate or 6-well plate, depending on sample size).

5. As with the cells in the 75-cm^2 flasks, cover the cells with complete medium and maintain in a 37°C environment with 5% CO_2. If the cells are being cultured for more than a few days, the medium should be replaced twice a week under sterile conditions.

6. *See* **Notes 4** and **8** to decide on your seeding and culture time protocols.

3.2. Immunohistochemistry (Fig. 1)

1. After the desired culture period (*see* **Note 4**), wash cells in PBS at 37°C (*see* **Note 5**).
2. Fix cells in 4% formaldehyde/PBS (with 2% added sucrose) for 15 min at 37°C (*see* **Note 5**).
3. Add permeablizing buffer and incubate for 5 min at 4°C (*see* **Note 6**).
4. Add 1% BSA/PBS at room temp and incubate for 5 min at 37°C.
5. Add primary antibody (for tubulin, vinculin, or vimentin 1:50 with 1%BSA/PBS) for 1 h at 37°C (can double stain for actin with phalloidin at this point) (*see* **Note 7**). If phalloidin is used to double stain with actin at this point, the samples should then be covered in foil to prevent light bleaching.
6. Wash three times for 5 min at room temperature in 0.5% Tween-20/PBS. Gentle agitation can be beneficial during the wash stages, e.g., using a slow plate shaker.
7. Add secondary, biotinylated antibody (1:50 with 1% BSA/PBS) for 1 h at 37°C.
8. Wash three times for 5 min at room temperature in 0.5% Tween-20/PBS.
9. Add either streptavidin-Texas red or streptavidin-FITC (1:50 with 1% BSA/PBS) for 30 min at 4°C (in fridge).
10. Wash three times for 5 min at room temperature in 0.5% Tween-20/PBS.
11. Mount with Vectorshield/DAPI to stain nucleus. Seal with nail polish. If you have cultured your cells on a material, a good idea is to mount your sample onto a large cover slip rather than bothering to attach it to a microscope slide. If you have cultured your cells on a cover slip, simply invert it onto a microscope slide.

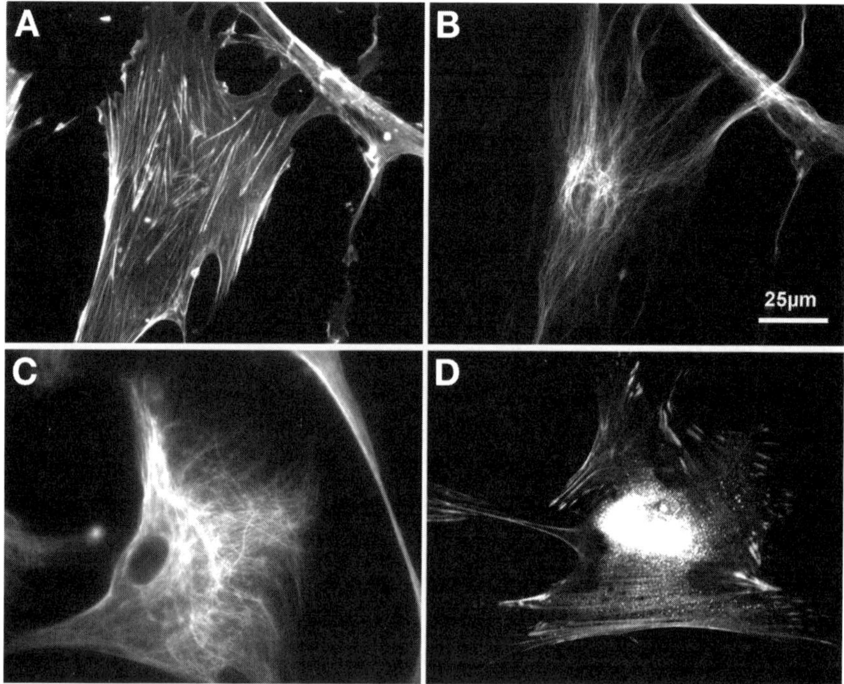

Fig. 1. Fluorescent images of cell cytoskeletons and focal adhesions. (**A**) Actin microfilaments stained with phalloidin–rhodamine. Note the large stress fibers throughout the cell. (**B**) Antibody-stained tubulin microtubules. Note the microtubule organizing center at the top of the nucleus. (**C**) Antibody-stained vimentin intermediate filaments. Note that the intermediate filaments completely surround the nucleus; they are the only cytoskeleton with direct linkage to the nucleus. (**D**) Antibody labeling of vinculin localized to focal adhesions. Note that vinculin is found throughout the cytoplasm and is enriched at the adhesions; thus you will get some level of background stain.

12. View by fluorescence microscopy. For a brief description of producing overlayed, color images see **Subheading 3.3.**

3.3. Producing Color Images in Photoshop®

1. Assuming that you have ended up with black-and-white images of actin, the nucleus, and either vimentin, tubulin, or vinculin of the same cell and now wish to make the color convergence image, open the three images you wish to merge in Adobe® Photoshop as 8-bit RGB images.
2. When on the nucleus image, select the channel menu (can be placed on desktop by selecting channel from the drop-down window menu).
3. Ensure that the background color is set to black (normally on the left-hand side of the screen).

4. Click on the green channel in the channel menu.
5. Select all (click on the select menu, then click all).
6. Cut the pixels (click on the edit menu, then click cut).
7. Repeat with the red channel.
8. Click on the actin image, select all, then click on the edit menu, then click copy.
9. Click on the nucleus image, select the red channel, then click on the edit menu, then click paste.
10. Repeat for either vimentin, tubulin, or vinculin, this time pasting into the green menu.
11. The brightness/contrast of each channel can be adjusted by clicking on the image menu, then the adjustments menu, then the brightness and contrast option. This should be adjusted to give the cleanest, brightest looking image without saturating the contrast.

3.4. Microarray (see Note 8)

3.4.1. RNA Extraction

1. Extract RNA using Qiagen RNeasy kit; make up additives for the buffers as described in the booklet (*see* **Note 9**).
2. Make up 600 µL of lysis buffer (mercaptoethanol [ME] and RLT) per sample—10 µL of ME should be added to each mL of buffer RLT.
3. Add 600 µL of lysis buffer with added ME to each sample; allow lysis to progress for 5–10 min. A cell scraper can be used to help with the lysis.
4. Dilute lysis buffer with 600 µL of 70% EtOH, and pipet into spin column (the column will take 700 µL, so you may have to divide into two runs of 600 µL). Centrifuge for 15 s at 7800*g* and discard the supernatant.
5. Add 700 µL of RW1 to the spin column. Centrifuge for 15 s at 7800*g* and discard the supernatant and collection tube.
6. Place the spin basket in a new tube, add 500 µL of RPE. Centrifuge for 15 s at 7800*g* and discard the supernatant.
7. Add 500 µL of RPE. Centrifuge for 2 min at 7800*g* and discard the supernatant and collection tube.
8. Place the spin basket into the capped tube provided.
9. Add 30 µL of dH$_2$O and centrifuge at 7800*g* for 1 min; then repeat so as to wash the membrane fully and give a final eluted volume of 60 µL. *Do not throw the water away after first wash—this is your extracted RNA.*
10. Work quickly because mRNA is unstable, and store RNA at −70°C.
11. Calculate amount of RNA extracted on a spectrophotometer at 260 nm using:

$$\text{RNA per mL} = (\text{absorbance reading}) \times 40 \times (\text{dilution})$$
(Note: Dilution is, e.g., 200 if 5 µL of RNA is diluted to 1000 µL.)

$$\text{Total RNA in sample} = (\text{RNA per mL}/1000) \times (\text{volume eluted})$$
(Note: Volume eluted refers to the amount of water used to wash RNA from spin column—should be 60 µL.)

You require at least 1 µg RNA for microarray—preferably 3–5 µg. For 3 µg:

$$3/(\text{total RNA}) \times (\text{volume eluted}) = \text{volume required for 3 µg RNA}$$

3.4.2. Reverse Transcription (see Note 10)

1. Keep all reagents on ice and combine the following per reaction (i.e., per sample; thus, if you have 5 sample replicates and 5 control replicates, this is 10 reactions):
   8.0 µL   5X first strand buffer
   1.5 µL   mRNA primer (12–18 mers)
   3.0 µL   20 m*M* dATP, dTTP, dGTP (*not* dCTP!)
   1.0 µL   2 m*M* dCTP
   1.0 µL   Cy 3 or Cy 5
   4.0 µL   0.1 *M* dithiothreitol
2. It is a good idea to make mastermixes for Cy 3 and Cy 5, i.e., calculate how much reagent you are going to need for Cy 3 and how much you are going to need for Cy 5. Prepare the two tubes with the reagents listed above, with one containing the Cy 3 and the other containing the Cy 5. Once the reagents are mixed, centrifuge for 5–10 s at 13,000*g* to ensure mixing.
3. If you are doing, e.g., 10 reactions, 5 should be with Cy 3 and the other 5 should be with Cy 5. The controls should be labeled the opposite of the samples. Fluor-flips should also be used, i.e., swap over the sample/control labeling. (**Note:** Keep dyes in the dark because they are light sensitive and very expensive.) For 10 reactions (5 control replicates and 5 test sample replicates):
   Sample   CY 3   CY 3   CY 3   CY 5   CY 5
   Control   CY 5   CY 5   CY 5   CY 3   CY 3
4. Divide the solutions into PCR tubes (18.5 µL of either Cy mix per tube, one tube per reaction). Add to each specific (and labeled) tube your 3 µg of RNA, and make the volume of each tube up to 40 µL with dH$_2$O. Centrifuge for 5 s at 7800*g* to ensure mixing.
5. Use thermocycler to incubate the PCR tubes at 65°C for 5 min, then at 42°C for 5 min.
6. Add 2 µL of Superscript II reverse transcriptase to each tube, centrifuge for 5 s to mix, and incubate at 42°C for 2–3 h.

3.4.3. Stop Reaction

1. Remove tubes from the thermocycler and put on ice.
2. Add 4 µL of 50 m*M* EDTA (pH 8.0) and 2 µL of 10 *N* NaOH to each tube.
3. Incubate the tubes in a thermocycler at 65°C for 20 min.
4. Remove from thermocycler and add 4 µL of 5 *M* acetic acid.
5. At this point add the sample and control reactions together in one tube so that you have both dyes in one tube (you should have now gotten rid of half of your tubes).
6. Add 100 µL of isopropanol to the combined tubes (*see* **Note 11**).
7. Precipitate on ice for 30 min, then centrifuge for 10 min at 4°C.
8. Pour off the isopropanol and rinse the pellet with ice-cold 70% EtOH (approx 50 µL —gently). *Do not dislodge the pellet.*

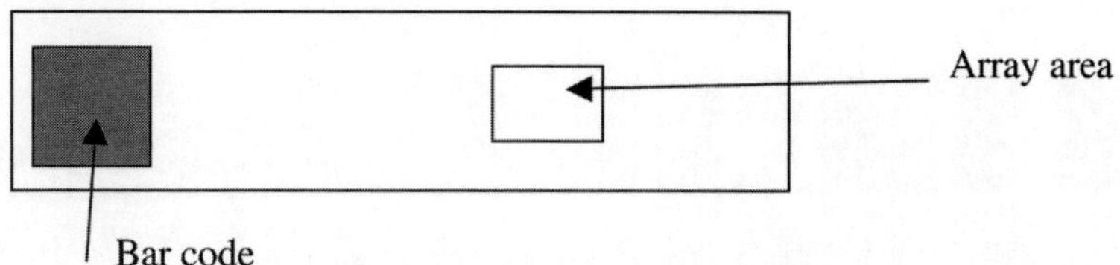

Fig. 2. The custom array slide.

9. Pour off the EtOH from each tube and gently blot the tube mouth on tissue. *Do not allow cDNA to dry completely.*
10. Resuspend the pellets in 5 µL of dH₂O (you should be able to note some color, but do not worry if you cannot).

3.4.4. Hybridization

1. Prepare hybridization solution (about 50 µL per slide—each tube goes on one slide). For each 100 µL of EasyHyb solution, add 5 µL of 10 mg/mL yeast tRNA and 5 µL of 10 mg/mL salmon sperm DNA.
2. Incubate the mixture at 65°C for 2 min and then cool to room temperature
3. Add 30 µL of the mixture to each PCR tube.
4. Pipet onto the slide on the array area (**Fig. 2**) and put on large coverslip. *Do not allow air bubbles to form.*
5. Place in hybridization chamber (*see* **Note 12**) containing a small amount of distilled water and *cover with foil.* Incubate at 37°C for 8–18 h (i.e., overnight).

3.4.5. Washing

1. Remove cover slips by quickly dipping in 1X SSC (let the cover slip slide off gently; do not prod or push if at all avoidable). Put slides into a staining rack and place in a staining dish with fresh 1X SSC.
2. Once all arrays have had cover slips removed, move into a staining dish containing 1X SSC with 0.1% sodium dodecyl sulfate (warmed to 50°C) three times for 10 min with gentle, occasional agitation.
3. After the washes, plunge the slides into 1X SSC four to six times.
4. Allow to dry by blotting (*do not touch array area*, rather touch the side of the slide onto tissue paper). Place in slide box and store in the dark, preferably with a desiccant to keep the slides dry.
5. Use a ScanArray Lite Microarray Analysis System (Packard Biochip Technologies, Billerica, MA), or a similar system, to compare red and green fluorescence and generate a .txt document that can be opened in spreadsheet programs for results export.

3.4.6. Cluster Analysis and Visualization of Microarray Data

1. Expression data are routinely analyzed by clustering genes into groups based on similarities in their expression profiles. This is done based on the methodologies developed by Eizen et al. *(16)*.

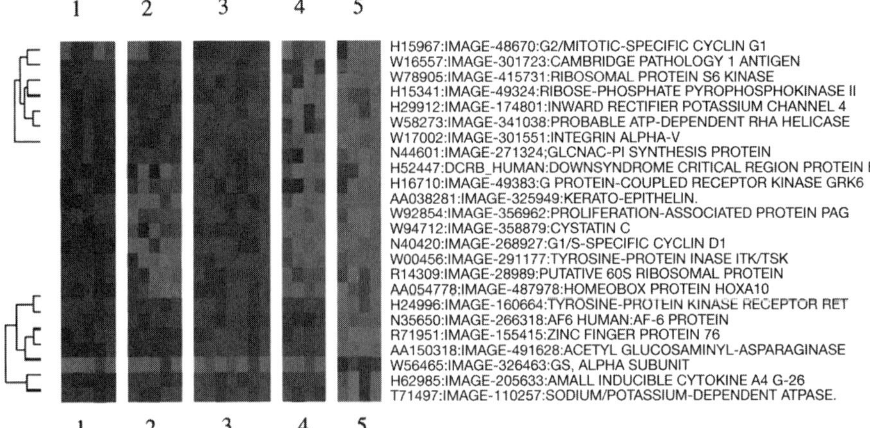

Fig. 3. Example of cluster analysis using TREEVIEW. HEK293 cells were attached to extracellular matrix components and treated with epidermal growth factor (EGF). Images are labeled with the corresponding sequence-verified gene names, IMAGE, and GenBank accession numbers and treatment identifiers: 1, poly-L-lysine + EGF; 2, fibronectin; 3, fibronectin + EGF; 4, laminin; and 5, laminin + EGF. For each treatment four to seven microarrays were used. Two subsets of genes are indicated with dendograms, which show similar responses to the treatments applied. (Reprinted with permission from **ref. *19*.**)

2. These clustering and display methods allow the detection of directional changes in gene expression profiles through multiple hybridizations of microarrays. The microarray data analysis package CLUSTER and TREEVIEW, developed in the Brown laboratory *(17,18)*, is used to group genes according to similar patterns of responsiveness. These programs are available to download as Freeware from the Eizen website (http://rana.lbl.gov/EizenSoftware.htm).

3. CLUSTER allows the organization of data, in the form of CY3-to-CY5 ratio values, from microarray hybridization for later analysis with TREEVIEW.

4. TREEVIEW allows the organized data to be visualized as a color-coded dendogram, with groups of gene identities clustered into blocks containing similar ratio values (**Fig. 3**).

5. These blocks are sorted on the basis of gene expression changes having similar magnitude and direction of change (either upregulated or downregulated). Usually genes that are upregulated are colored red and those that are downregulated are colored green (**Fig. 3**).

6. First, data text files obtained from the ScanArray Lite Microarray Analysis System should be imported into CLUSTER as tab-delimited text files in the format described in the HELP function of the program. In the input tables, rows represent genes and columns represent samples or observations (e.g., a single microarray hybridization). These files can be created and managed using Microsoft Excel.

7. Once imported into CLUSTER the data should then be log transformed to base 2. For example, the raw ratio values 1.0, 2.0, and 0.5 become 0, 1.0, and -1.0 when log transformed. With these values, twofold up and twofold down are symmetrical about 0. This allows the fluorescent ratio values from the microarray analysis to have directionality, which is essential for TREEVIEW to color-code the direction of gene expression changes (either upregulated, positive value or downregulated, negative value).

8. To deal with the effects of experimental variation between array experiments (e.g., differences in labeling efficiencies or small variations in the surface of individual arrays), CLUSTER allows centering of log-transformed ratio values. This should be done next. Eizen recommends using median rather than mean centering, and we do this routinely. Here, means from each column (with a single column of data representing all the ratio values from a single hybridized microarray) is subtracted from the values in each column of data, so that the median value of each column is 0.

9. Next, CLUSTER allows the normalization of each data column. Normalization sets the magnitude (sum of the squares of the values) of each column to 1.0. This is used to standardize the two fluorescence intensity channels per microarray.

10. Following this, the data set can be subjected to hierarchial clustering. This allows the arrangement of gene identities and the corresponding ratio values from each array experiment to be assembled into a tree (dendogram), with short branches for very similar values and longer branches if values are dissimilar.

11. After clustering CLUSTER produces a .cdt (clustered data table) file, which appears on your desktop. This should be opened with TREEVIEW to view the resulting dendogram. Subgroups or blocks of similar changes can be highlighted and downloaded as a separated file for presentation of results (**Fig. 3**).

4. Notes

1. It is recommended that for sterility, the complete media is aliquotted into 50-mL sterile pots for freezing before use. If the media is cloudy on defrosting, we recommend filtering to prevent cloudiness.

2. Aliquot into 20-µL amounts in Eppendorf tubes and store at –70°C. All the antibodies are used at a 1:50 dilution, and thus 20 µL can be reconstituted in 1 mL of PBS/BSA for staining.

3. All solutions are fine to use until they go cloudy. The BSA/PBS will go off within a week, so only the amount needed should be prepared.

4. It is best to view the cells for cytoskeleton and adhesions when they are separated from one another; thus, we recommend a seeding density of 1×10^4 cell/mL and culture for 3–4 d as gives you best results. This has to be carefully thought out when considering microarray (*see* **Note 8**).

5. The cytoskeleton, notably intermediate filaments, is temperature sensitive; thus the PBS and fixative should be warmed to 37°C prior to use.

6. PBS/BSA, permeablizing buffer, PBS/Tween-20 can be used at room temperature. This includes antibody dilutions in PBS/BSA.

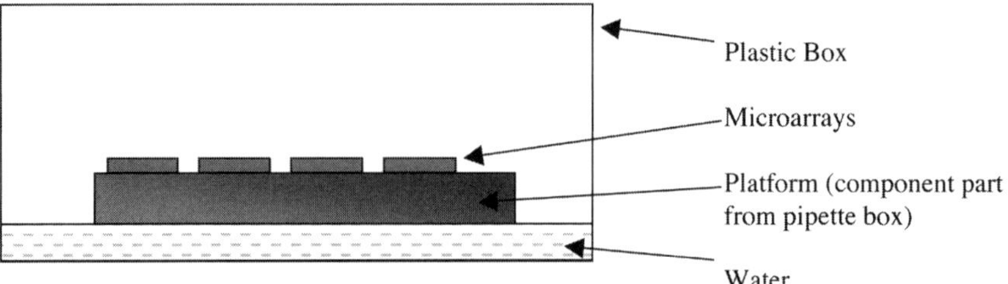

Fig. 4. The hybridization chamber

7. We have recommended that the antibodies be detected with FITC (green fluorescence) and actin labeled with rhodamine (via phalloidin) (red fluorescence). This can be swapped as long as they are labeled with different colors! To make up the antibody/phalloidin dilution, add 20 µL of phalloidin and 20 µL of primary antibody to 1 mL of PBS/BSA.

8. First, if you have more than one treatment, each treatment should be considered a separate experiment. The procedure outlined below is for one experiment and should be expanded to suit your needs. Second, you must have the same number of samples and controls, e.g., five sample replicates and five control replicates. If you are doing more than one experiment (or treatment), you only need to have one set of controls provided you have sufficient mRNA, e.g., five replicates for test 1, five replicates for test 2, and five control replicates. Last, you need to have sufficient cell numbers for mRNA extraction. This can be tricky if you are trying to directly compare cell densities used for immunofluorescence and microarray—you will need to extract from very large areas for microarray, e.g., a whole 75-cm^2 flask during a proliferative growth phase for the cells. If it is not so important that the time points are the same, grow to around 90% confluency on an area of at least 2 cm^2. If this is not possible, consider combining multiple extractions in one spin column, e.g., for 5 microarray replicates use 10 treatment replicates and pass lysate for 2 sample down 1 spin column.

9. ME—*open only in fume cupboard or class II hood*—should only be added to buffer RLT immediately before extraction; also, only use the volumes of each that you require.

10. You are adding very small quantities of samples to the PCR tubes. In order to ensure mixing and that small droplets are not stuck to the tube walls, at each step, once you have added all the reagents, give the tubes a quick centrifuge (2–3 s at 7800g).

11. After a few steps you will need a refrigerated microfuge—if you do not have one, put the microfuge in a cold room at this time.

12. A hybridization chamber may simply be a plastic lunchbox (airtight) with a stage (e.g., from a pipet rack box) and a small amount of water in the bottom (**Fig. 4**).

Acknowledgments

M.D. is a BBSRC David Phillips Fellow and is funded through this route.

References

1. Dalby, M. J., Yarwood, S. J., Riehle, M. O., Johnstone, H. J., Affrossman, S., and Curtis, A. S. (2002) Increasing fibroblast response to materials using nanotopography: morphological and genetic measurements of cell response to 13-nm-high polymer demixed islands. *Exp. Cell Res.* **276,** 1–9.
2. Dalby, M. J., Riehle, M. O., Yarwood, S. J., Wilkinson, C. D., and Curtis. A. S. (2003) Nucleus alignment and cell signaling in fibroblasts: response to a microgrooved topography. *Exp. Cell Res.* **284,** 274–282.
3. Burridge, K. and Chrzanowska-Wodnicka, M. (1996) Focal adhesions, contractility, and signaling. *Annu. Rev. Cell Dev. Biol.* **12,** 463–518.
4. Knight, M. M., van de Breevaart Bravenboer, J., Lee, D. A., van Osch, G. J., Weinans, H., and Bader, D. L. (2002) Cell and nucleus deformation in compressed chondrocyte-alginate constructs: temporal changes and calculation of cell modulus. *Biochim. Biophys. Acta* **1570,** 1–8.
5. Dahl, K. N., Kahn, S. M., Wilson, K. L., and Discher, D. E. (2004) The nuclear envelope lamina network has elasticity and a compressibility limit suggestive of a molecular shock absorber. *J. Cell Sci.* **117,** 4779–4786.
6. Bloom, S., Lockard, V. G., and Bloom, M. (1996) Intermediate filament-mediated stretch-induced changes in chromatin: a hypothesis for growth initiation in cardiac myocytes. *J. Mol. Cell. Cardiol.* **28,** 2123–2127.
7. Heslop-Harrison, J. S. (1992) Nuclear architecture in plants. *Curr. Opin. Genet. Dev.* **2,** 913–917.
8. Heslop-Harrison, J. S., Leitch, A. R., and Schwarzacher, T. (1993) The physical organization of interphase nuclei, in *The Chromosome* (Heslop-Harrison, J. S., et al., eds.), Bios, Oxford, pp. 221–232.
9. Heslop-Harrison, J. S. (2003) Planning for remodelling: nuclear architecture, chromatin and chromosomes. *Trends Plant Sci.* **8,** 195–197.
10. Dalby, M. J., Riehle, M. O., Sutherland, D. S., Agheli, H., and Curtis, A. S. (2004) Use of nanotopography to study mechanotransduction in fibroblasts—methods and perspectives. *Eur. J. Cell Biol.* **83,** 159–169.
11. Juliano, R. L. and Haskill, S. (1993) Signal transduction from the extracellular matrix. *J. Cell Biol.* **120,** 577–585.
12. Vuori, K. (1998) Integrin signaling: tyrosine phosphorylation events in focal adhesions. *J. Membr. Biol.* **165,** 191–199.
13. Cary, L. A., Han, D. C., and Guan, J. L. (1999) Integrin-mediated signal transduction pathways. *Histol. Histopathol.* **14,** 1001–1009.
14. Cary, L. A. and Guan, J. L. (1999) Focal adhesion kinase in integrin-mediated signaling. *Front. Biosci.* **4,** D102–D113.
15. Bershadsky, A. D., Balaban, N. Q., and Geiger, B. (2003) Adhesion-dependent cell mechanosensitivity. *Annu. Rev. Cell Dev. Biol.* **19,** 677–695.
16. Eizen, M. B., Spellman, P. T., Brown, P. O., and Botstein, D. (1998) Cluster analysis and display of genome-wide expression patterns. *Proc. Natl. Acad. Sci. USA* **95,** 14,863–14,868.

17. Schena, M., Shalon, D., Heller, R., Chai, A., Brown, P. O., and Davis, R. W. (1996) Parallel human genome analysis: microarray-based expression monitoring of 1000 genes. *Proc. Natl. Acad. Sci. USA* **93,** 10,614–10,619.
18. Shalon, D., Smith, S. J., and Brown, P. O. (1996) A DNA microarray system for analyzing complex DNA samples using two-color fluorescent probe hybridization. *Genome Res.* **6,** 639–645.
19. Yarwood, S. J. and Woodgett, J. R. (2001) Extracellular matrix composition determines the transcriptional response to epidermal growth factor receptor activation. *Proc. Natl. Acad. Sci. USA* **98,** 4472–4477.

11

Proteomic Analysis of Cell Surface Membrane Proteins in Leukemic Cells

Robert S. Boyd, Martin J. S. Dyer, and Kelvin Cain

Summary

Plasma membrane proteins play a key role in cellular processes such as migration, adhesion, and cell survival. The comprehensive annotation of the leukemic cell plasma membrane proteome allows the identification of proteins that may be involved in the pathogenesis of disease and may provide novel therapeutic targets. The identification of known adhesion molecules or novel proteins with similar attributes to adhesion molecules provides the starting point for the generation of hypothesis on the role of these proteins in adhesion processes. In order to identify these novel proteins, we have developed a proteomics methodology using purified plasma membranes prepared from human leukemic cells.

Key Words: Proteomics; plasma membrane; cell surface; leukemia; mass spectrometry.

1. Introduction

The plasma membrane contains many proteins known to be overexpressed in disease processes. These proteins may be utilized either for diagnosis or for therapy. To be able to comprehensively define the set of proteins expressed in this subcellular compartment would also shed light on disease pathogenesis. The proteomic analysis of the plasma membrane protein composition of a limited number of patient samples coupled with a data-mining strategy allows the identification of a subset of proteins of biological interest because of their known function or altered expression in disease states. The relative expression levels of these proteins in normal and disease samples can then be profiled across a much larger number of samples using techniques such as quantitative reverse transcription–polymerase chain reaction or Western blotting. This approach can target proteins that may be intimately involved in the generation of the disease state and provides the possibility of rationale-based therapeutic intervention.

From: *Methods in Molecular Biology, vol. 370:*
Adhesion Protein Protocols, Second Edition
Edited by: A. S. Coutts © Humana Press Inc., Totowa, NJ

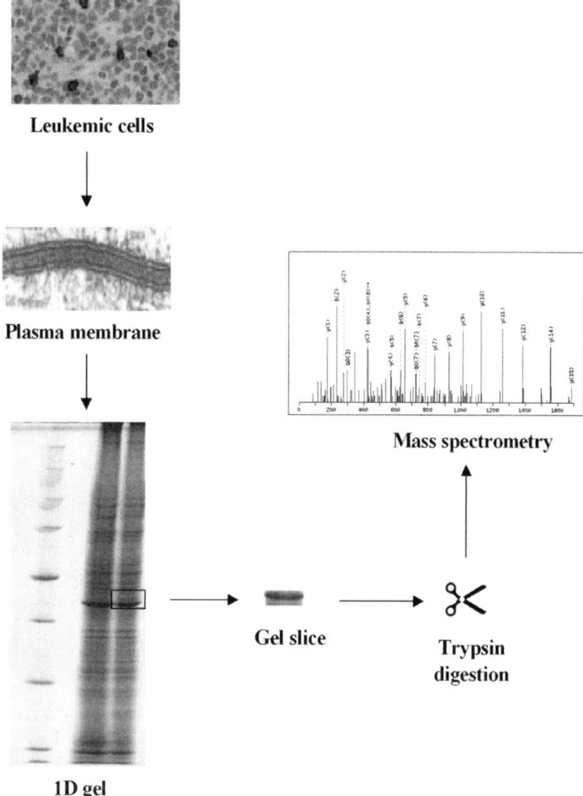

Fig. 1. Flow path for a typical experiment.

The analysis of membrane proteins by conventional two-dimensional gel electrophoresis technology and mass spectrometry is hampered by problems of solubility, and we have therefore developed an alternative method for analyzing plasma membranes. In this approach (**Fig. 1**), leukemic cells were purified from patient blood samples and disrupted using a ball-bearing homogenizer. The cell membranes were harvested by centrifugation and purified on a preformed 15–60% (w/v) sucrose gradient at 100,000*g* for 17 h. Proteins from the fractionated sucrose gradient were then analyzed by sodium dodecyl sulfate–polyacrylamide gel electrophoresis (SDS-PAGE) and Western blotting for known plasma membrane and other cellular markers. Those fractions exhibiting positive immunoreactivity for plasma membrane proteins but little or no immunoreactivity for other organelle markers were used as the purified plasma membrane preparation. Purified plasma membrane fractions were then separated by SDS-PAGE and the gel lanes sliced into 1.5-mm slices, which were subjected to tryptic digestion *(1)*. Because each gel slice contains many proteins, proteolytic digestion produces a complex peptide mixture, which must be separated by microcapillary high-performance liquid chromatography (HPLC). The column eluate was fed directly into a mass spectrometer through a nanospray ion source to generate ionized peptides. These were analyzed for their mass-to-charge ratio intensities, and selected peptides were fragmented/sequenced by tandem mass spectrometry (MS/MS). The sequence spectra were then searched against public domain databases using a search engine such as Mascot (Matrix Science, London) or Sequest (Thermo, MA).

This approach can identify many different proteins in a single gel slice, and it has been successfully used to identify novel proteins such BCNP1, MIG2B, and other proteins in the plasma membrane of leukemic cells obtained from patients with chronic lympocytic leukemia *(2)*. This proteomic methodology has also been used in a comprehensive analysis of breast cancer cell membranes to detect known and unique proteins, which are associated with clinical cancer expression *(3)*. Thus, this proteomics methodology offers an innovative strategy for identifying novel membrane proteins whose protein expression is altered through disease or pathological insult.

2. Materials

2.1. Preparation of Leukemic B-Cells

1. RPMI-1640 (Gibco, Invitrogen, Carlsbad, CA).
2. Histopaque-1077 (Sigma, St. Louis, MO).
3. Lithium heparin tubes (Becton-Dickinson, Franklin Lakes, NJ).
4. Trypan blue (Sigma).
5. Phosphate-buffered saline: 6.4 mM Na$_2$HPO$_4$, 1.47 mM NaH$_2$PO$_4$, 130 mM NaCl, 2.68 mM KCl pH 7.4.

2.2. Preparation of Plasma Membranes

1. Homogenization buffer: 250 mM sucrose, 10 mM N-2-hydroxyethylpiperazine-N'-2-ethanesulfonic acid (HEPES), pH 7.4, 1 mM ethylene diamine tetraacetic acid (EDTA), and 0.02% (w/v) sodium azide stored at 4°C (*see* **Note 2**).
2. Ball-bearing homogenizer (Isobiotec, Heidelberg, Germany) with 18-μm clearance ball (*see* **Note 3**).
3. 19- and 25-gage needles (Sigma).
4. Sucrose buffer: 60% sucrose (w/v), 10 mM HEPES, pH 7.4, 1 mM EDTA, and 0.02% (w/v) sodium azide, stored at 4°C. Use a refractometer to determine the sucrose content of buffers (*see* **Note 4**).
5. Protease inhibitor cocktail (mammalian) (Sigma). The cocktail is stable for greater than 1 yr when stored at −20°C.

6. Polyallomer (14 × 95 mm) centrifuge tubes (Beckman Coulter, Fullerton, CA).
7. Bradford protein assay (Bio-Rad, Hercules, CA).

2.3. Western Blotting for Organelle Markers

1. One-dimensional sample buffer 4X: 250 m*M* Tris-HCl, pH 6.8, 8% (w/v) SDS, 40% (v/v) glycerol, 20% (v/v) 2-mercaptoethanol, 0.001% (w/v) bromophenol blue. Store at room temperature.
2. 4–20% Acrylamide, 1 mm, Tris-glycine minigels (Invitrogen).
3. Running buffer: 25 m*M* Tris, 190 m*M* glycine, 1% (w/v) SDS.
4. Precision prestained molecular-weight markers (Bio-Rad).
5. Hybond-C extranitrocellulose membrane (Amersham, Chiltern Hills, UK).
6. Transfer buffer: 25 m*M* Tris-HCl, 190 m*M* glycine 20% (v/v) methanol.
7. TBST buffer: 10 m*M* Tris-HCl, pH 7.6, 150 m*M* NaCl, and 0.05% (v/v) Tween-20.
8. Blocking buffer: 5% (w/v) milk powder in TBST.
9. Primary antibodies: mouse anti-OxPhos II 70-kDa subunit (Molecular Probes, OR; use at 1:1000 dilution), mouse anti-calnexin (Affinity Bioreagents, Golden, CO; use at 1:2000 dilution), mouse anti-mitochondrial heat-shock protein 70 (Affinity Bioreagents; use at 1:1000 dilution), mouse anti-transferrin receptor (Zymed Laboratories Inc., South San Francisco, CA; use at 1:500 dilution), and mouse anti-CD20 (Santa Cruz Biotechnology, Santa Cruz, CA; use at 1:1000 dilution).
10. Secondary antibody: anti-mouse IgG conjugated to horseradish peroxidase (Amersham).
11. Enhanced chemiluminescent Western blotting detection reagents (Amersham).

2.4. SDS-PAGE for Mass Spectrometry

1. Mark 12 unstained molecular-weight markers (Invitrogen).
2. Brilliant blue G-collidal concentrate (Sigma).
3. Montage in-gel proteolysis kit (In-gel Digest ZP kit) (Millipore, Billerica, MA).
4. Gel fixing buffer: 7% (v/v) acetic acid in 40% (v/v) methanol.
5. Brilliant blue G-colloidal staining solution (Sigma; *see* **Note 5**).
6. Gel destaining solution: 10% (v/v) acetic acid in 25% (v/v) methanol.

2.5. Membrane Protein Digestion

1. Destain buffer 1: 25 m*M* ammonium bicarbonate and 5% (v/v) acetonitrile.
2. Destain buffer 2: 25 m*M* ammonium bicarbonate and 50% (v/v) acetonitrile.
3. Trypsin resuspension buffer: 1 m*M* HCl.
4. Trypsin buffer: 25 m*M* ammonium bicarbonate.
5. Extraction buffer: 0.2% (v/v) trifluoroacetic acid (Sigma).
6. Elution buffer: 0.1% trifluoroacetic acid (v/v) and 50% (v/v) acetonitrile.

2.6. Mass Spectrometry of Membrane Protein Peptides

1. Analytical coloumn: Atlantis™ dC$_{18}$, 3 μm (Waters, Milford, MA)
2. Trap column: Atlantis dC$_{18}$, 5 μm OPTI-PAK™ (Waters).

3. Buffer A: 0.1% (v/v) formic acid, 5% (v/v) acetonitrile.
4. Buffer B: 95% (v/v) acetonitrile, 0.1% (v/v) formic acid.
5. Buffer C: 0.1% (v/v) formic acid.

3. Methods

3.1. Preparation of Purified Leukemia Cells

1. Select patients with lymphocyte counts of greater than 50×10^6/mL to minimize the possibility of contamination of tumor cells by normal hematopoietic cells (*see* **Note 6**).
2. Collect blood samples into lithium heparin tubes.
3. Add 15 mL of Histopaque-1077 to a 50-mL tube. Dilute blood (15 mL) with an equal volume of RPMI-1640 media and carefully layer on top of the Histopaque-1077 using a plastic Pasteur pipet.
4. Spin tube at 400g for 30 min at room temperature.
5. Collect the white blood cells at the interphase between the Ficoll-Isopaque medium and the blood media using a Pasteur pipet.
6. Wash leukemic cells twice in 50 mL of RPMI-1640 media by centrifugation at 800g for 10 min. Count cells using a hematocytometer and stain with Trypan blue (0.125% w/v) in phosphate-buffered saline to assess their viability (*see* **Note 7**).
7. Aliquot leukemic cells into 0.5×10^9 lots and pellet by centrifugation at 800g for 10 min. Freeze the cells quickly on dry ice and store at −80°C.

3.2. Preparation of Plasma Membranes

1. Add 0.5 mL of protease inhibitor cocktail to 20 mL of ice-cold homogenization buffer.
2. Thaw cell pellets (1–2×10^9) quickly and resuspended in homogenization buffer + protease inhibitor cocktail (20 mL of homogenization buffer per 1×10^9 cells).
3. Generate a single cell suspension by passing the cells through a 19-gage needle three times and a 23-gage needle three times (*see* **Note 8**).
4. Place a 5-mL syringe on either end of the prechilled ball-bearing homogenizer and pass the cells through the chamber 10–15 times using a ball bearing with an 18-µm clearance. (Cells are processed in 5-mL aliquots.) Examine the resulting homogenate using phase contrast microscopy every three to five cycles (*see* **Note 9**).
5. Centrifuge the cell homogenate at 1000g for 10 min at 4°C to pellet intact cells and nuclei.
6. Centrifuge the postnuclear cytosol (PNS) at 3000g for 10 min at 4°C.
7. Layer the PNS onto a 2-mL 60% (w/v) sucrose cushion in a polyallomer (14×95 mm) centrifuge tube and centrifuge the tube in a SW 40 Ti rotor at 100,000g for 1 h with a slow acceleration and deceleration.
8. Harvest the crude plasma membrane from the top of the sucrose cushion using a 1-mL Pasteur pipet.
9. Measure the refractive index of the PNS and adjust with 10 mM HEPES buffer, pH 7.4, to give a sucrose content of less than 15% (w/v).

10. Prepare a 15–60% (w/v) sucrose gradient in a polyallomer (14 × 95 mm) centrifuge tube and layer 2 mL of the PNS on top of the gradient. Centrifuge the tube in a SW 40 Ti rotor at 100,000g for 17 h with a slow acceleration and deceleration (*see* **Note 10**).
11. Fractionate the sucrose gradient in 0.5-mL steps from the top of the tube with a hooked Pasteur pipet (*see* **Note 11**).
12. Assay the protein content of the fractions using the Bradford protein assay.

3.3. Western Blotting for Organelle Markers

These instructions relate to the Invitrogen Xcell *Surelock*™ Mini-Cell system and the use of precast acrylamide (4–20%) Tris-glycine minigels. The use of precast minigels gels helps to reduce any source of variability during proteomic analysis of membrane proteins. The key steps, however, are similar to those with any minigel system.

1. Set up the minigel system according to the manufacturer's instructions and add 500 mL of running buffer to the upper and lower chambers of the tank.
2. Wash out the sample wells with a Pasteur pipet and load the prestained molecular weight markers (10 μL) and plasma membrane fraction. (5–10 μg of protein).
3. Electrophorese the gels for 2 h (25 mA/gel) and remove them from the plastic cassette.
4. Soak precut sheets of 3MM paper and nitrocellulose sheets in transfer buffer for 5 min.
5. Place two sheets of 3MM paper on one side of the blotting cassette and position the gel on top of the 3MM paper. Place the nitrocellulose carefully on the gel, ensuring that no air bubbles are trapped between the two surfaces. Overlay a further two 3MM sheets on the nitrocellulose sheet.
6. Place the blotting cassette in the gel tank with (important) the nitrocellulose sheet facing the anode and add approx 700 mL of transfer buffer. Insert a frozen Bio-Ice cooling unit into the buffer chamber to ensure that the tank temperature remains between 10 and 15°C.
7. Circulate the tank buffer with a magnetic stirrer, the speed of which has been adjusted to ensure adequate cooling of the buffer.
8. Place the lid on the tank and run the blotting cassette at 50 V for 2 h.
9. Remove the blot from the tank and incubate for 1 h in blocking buffer on a rocking platform.
10. Add 10 mL of primary antibody in 2% milk powder to the blot for 1 h.
11. Decant the primary antibody and wash the blot three times (1 min) with TBST.
12. Add 10 mL of secondary antibody in 2% milk powder to the blot for 1 h.
13. Decant the secondary antibody and wash the blot four times (10 min) in TBST.
14. Mix 2.5-mL aliquots of prewarmed enhanced chemiluminescent reagent A and B and add to the blot for 1 min with gentle rocking. Remove excess reagent from the blot by blotting with tissue and seal the blot in saran wrap.

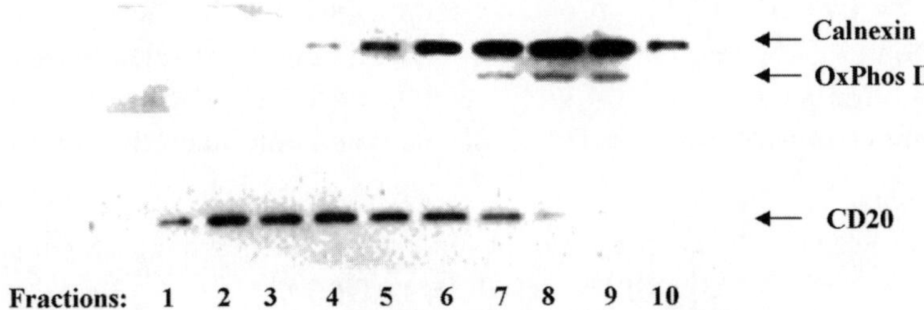

Fig 2. CD20, calnexin, and oxidoreductase II immunoreactivity in sucrose gradient membrane protein fractions. Progressive B-CLL membrane fractions (1–10) were analyzed by immunoblotting with anti-calnexin, oxidoreductase II, and anti-CD20 antibody.

15. Place sealed blot inside an X-ray cassette with film for 1–10 min. Develop film in an automatic film processor. A typical result is shown in **Fig. 2.**

3.4. SDS-PAGE for Mass Spectrometry

1. Pellet the plasma membranes by diluting the sucrose in the membrane fraction with 10 mM HEPES, pH 7.4 to less than 5% (w/v) and spin the sample at 100,000g for 1 h in a micro-ultracentrifuge (*see* **Note 12**).
2. Solubulize the membrane pellet in one-dimensional lysis buffer to give a protein concentration ≥ 3 µg/µL.
3. Run a precast gel as described in **Subheading 3.3.** Load the gel with unstained protein markers and 30–40 µg of purified plasma membrane fraction.
4. Incubate the gel in fixing buffer for 1 h on a rocking platform.
5. Add 40 mL of brilliant blue G-colloidal working solution to 10 mL of methanol to make up 50 mL of brilliant blue G-colloidal staining solution. Mix this solution vigorously for 30 s.
6. Decant the fixing buffer and add 50 mL of brilliant blue G-colloidal staining solution to the gel. Incubate the gel in this solution for 2 h on a rocking platform.
7. Decant the staining solution and add 50 mL of destaining buffer to the gel for 40 s with gentle shaking.
8. Quickly decant the destaining solution and rinse the gel with 50 mL of 25% methanol. Destain the gel overnight in 25% methanol, and store in the dark at 4°C.

3.5. Membrane Protein Digestion

The following protocol describes the use of the Millipore Montage In-Gel Digest$_{ZP}$ kit for proteolysis of protein in gel slices; the protocol is adapted from a published method *(4)*. The key component of the kit is a zip plate, a 96-well microtiter plate with 300 nL of C$_{18}$ media immobilized at the bottom of each well. The plate is connected to a vacuum manifold and has two main advantages over the use of individual eppendorf tubes: (1) 96 gel slices can be processed relatively quickly in one procedure, and (2) the incorporation of a small amount of C_{18} resin in the base of the well allows the selective elution of tryptic peptides into collection wells, which eliminates the need to clean up tryptic peptide digests using ZipTips (Millipore) and minimizes the elution buffer volume.

1. Using a clean scalpel, cut a series of 3×1.5 mm slices from each gel lane and transfer the slices to the zip plate wells (*see* **Note 13**).
2. Add a gel slice containing 100 ng of bovine serum albumin to one of the wells to be used as a quality control standard.
3. Incubate the gel slices for 30 min in 150 µL of destain buffer 1.
4. Apply full vacuum to the plate (15 mmHg) to remove the destain buffer and add 150 µL of destain buffer 2 to each well. Incubate the gel pieces for a further 30 min, and use full vacuum to remove buffer 2 from the wells.
5. Repeat **step 4**.
6. Dehydrate the gel pieces by incubation in 200 µL of 100% acetonitrile for 10 min.
7. Apply full vacuum for 2 min to remove all acetronitrile from the wells.
8. Hydrate the gel pieces by the addition of 15 µL of trypsin solution (*see* **Note 14**). Seal the wells and incubate the gel pieces for 3 h at 37°C.
9. Add 8 µL of acetonitrile directly to the C_{18} resin in each well and incubate the plate for 15 min at 37°C (*see* **Note 15**).
10. Incubate each well for 30 min with 130 µL of the extraction buffer.
11. Remove the extraction buffer from the wells by application of partial vacuum pressure (7 mmHg). Wash the wells once with 100 µL of extraction buffer and remove this buffer by the application of full vacuum pressure for 2 min (*see* **Note 16**).
12. Remove the gel pieces from the wells and add 30 µL of elution buffer to the wells.
13. Apply partial vacuum pressure (7 mmHg) to the wells to elute the tryptic peptides from the C_{18} resin. Collect the eluates in a 96-well collection tray placed under the zip plate.
14. Transfer the elution buffer in each well to a series of 650-µL Eppendorf tubes. Place these tubes inside 1.5-mL screw cap Eppendorf tubes and spin in a speedvac until dry (*see* **Note 17**).
15. Solubulize the dried-down tryptic peptides in 5 µL of 5% formic acid (*see* **Note 18**).

3.6. Mass Spectrometry of Membrane Tryptic Peptides

This section assumes some familiarity with nanospray as an ionization technique for tryptic peptides, MS/MS, analysis and the use of microscale capillary HPLC. In this study we have used a quadruple time-of-flight mass spectrometer (Waters) coupled to a nanoLC system (CapLC; Waters) via a nanospray interface (*see* **Fig. 3**), although there are alternative nanoLC MS/MS systems that are capable of performing MS/MS analysis for this type of study.

1. Aliquot 6 µL of peptide digest into a 0.1-mL polypropylene vial (Kinesis, Bedfordshire, UK).

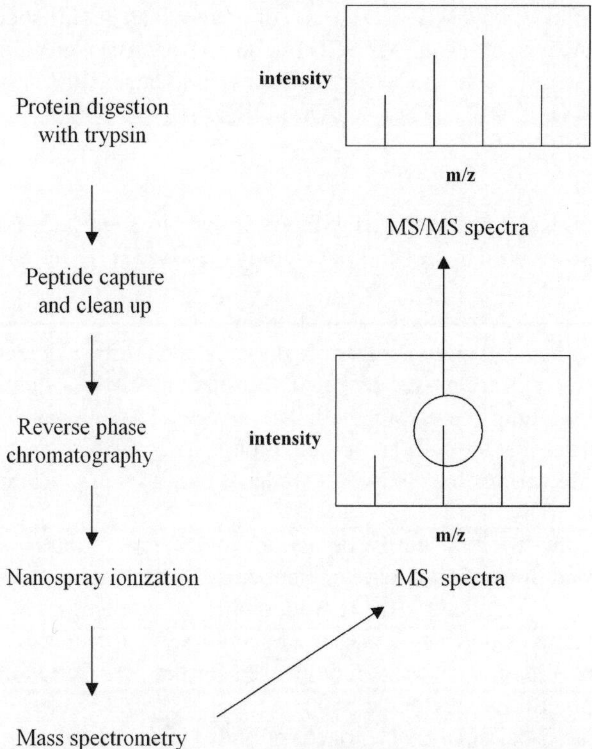

Fig 3. Peptide sequencing by tandem mass spectrometry.

2. Set the CapLC autosampler to µL pick-up mode to take up 5 µL of peptide digest.
3. Preconcentrate the peptide mixture by loading (flow rate of 30 µL/min) onto a trap column in 0.1% formic acid.
4. Elute the peptides from the trap column at a flow rate of 200 nL/min onto a C_{18} analytical column by a linear gradient of acetonitrile (7 to 80% acetonitrile over 50 min). This linear acetonitrile gradient provided the basis of chromatographic reverse-phase separation.
5. The CapLC was interfaced to the Q-ToF using a nanospray Z source.
6. Acquire MS and MS/MS data using an automatic data acquisition program. This program generates fragmentation or MS/MS spectra at selected mass-to-charge ratio values, which are determined from the MS analysis, primarily according to the criterion of ion signal intensity.
7. Process the MS data using MassLynx 4.0 software and carry out database searching using the MASCOT program (Matrix Science). The parameters used for Mascot searches were as follows: three missed cleavages, 0.4-Dalton mass accuracy for the parent and fragment ions, respectively, and variable oxidization of the methionine. Proteins identified by two peptides with MASCOT scores of more than 40 were accepted without manual validation. In cases where the protein was identified by

a single peptide with a MASCOT score of greater than 30, the spectra were interpreted manually to check the MASCOT result. All database searching used the nonredundant SwissProt database (ftp://us.expasy.org/databases/). A typical result for the membrane protein CD22 is shown in **Fig. 4**.

4. Notes

1. All solutions should be prepared with water that has a resistance of 18.2 MΩ-cm For all solutions used in the MS system, only degassed HPLC-grade water (Sigma) should be used.
2. Thimerosal is a less toxic antibacterial agent.
3. The ball-bearing cell homogenizer breaks cells open in an efficient and reproducible manner. Cells are passed repeatedly through a bore in a stainless steel block that contains a tungsten carbide ball. The narrow clearance between the ball and the bore means that the cells are forced through a narrow passage that breaks them up. There are various ball sizes, but we have found 18-µm clearance optimal for lymphocyte lysis.
4. Solutions with different sucrose content have different refractive indexes. Under standard conditions a refractometer can be used to determine the sucrose content of solutions; it is crucial for the reproducibility of gradient fractionation that the sucrose content of buffers be measured using a refractometer. Protease inhibitor must be added to all sucrose solutions used for the purification of plasma membranes (1:50 dilution of stock).
5. Brilliant blue G-colloidal concentrate is provided in a solid form. Eight hundred mL of deionized water are added to the bottle, and the contents are mixed by inversion. This is the working stock.
6. We have purified both mantle and chronic lymphocytic leukemia cells using this method.
7. It is important to fully characterize the purified leukemic cells; e.g., the quality and purity of chronic lymphocytic leukemic cells were assessed by fluorescence-activated cell sorting using anti-CD19 and anti-CD5 antibodies.
8. Care must be taken in using needles with human cells. A single cell suspension is important for the action of the ball-bearing homogenizer.
9. It is vital to monitor the cell lysis process using a phase contrast microscope. The number of homogenization cycles, which breaks open most cells but does not disrupt cell organelles, is the optimal number for pure plasma membrane generation.
10. A 15–60% (w/v) sucrose gradient can be generated through a gradient mixer (Bio-Comp Instruments, Fredericton, Canada). We generated a sucrose gradient using an empirical method in which 1-mL aliquots of sucrose solution were carefully aliquoted on top of each other (60%, 54%, 50%, 46%, 42%, 38%, 34%, 30%, 26%, 22%, 18%, 15%) in a 14-mL polyallomer centrifuge tube; the tube was then incubated for a minimum of 6 h to allow a linear gradient to form. It is important to determine with a refractometer that a reproducible gradient is formed under the researcher's own conditions. The use of a SW 40 Ti or SW 60 rotor depends on

A

```
  1  MHLLGPWLLL LVLEYLAFSD SSKWVFEHPE TLYAWEGACV WIPCTYRALD
 51  GDLESFILFH NPEYNKNTSK FDGTRLYEST KDGKVPSEQK RVQFLGDKNK
101  NCTLSIHPVH LNDSGQLGLR MESKTEKWME RIHLNVSERP FPPHIQLPPE
151  IQESQEVTLT CLLNFSCYGY PIQLQWLLEG VPMR**QAAVTS TSLTIK**SVFT
201  RSELKFSPQW SHHGK**IVTCQ LQDADGK**FLS NDTVQLNVKH TPKLEIK**VTP**
251  **SDAIVR**EGDS VTMTCEVSSS NPEYTTVSWL KDGTSLKK**QN TFTLNLR**EVT
301  KDQSGKYCCQ VSNDVGPGRS EEVFLQVQYA PEPSTVQILH SPAVEGSQVE
351  FLCMSLANPL PTNYTWYHNG KEMQGRTEEK VHIPKILPWH AGTYSCVAEN
401  ILGTGQRGPG AELDVQYPPK **KVTTVIQNPM PIR**EGDTVTL SCNYNSSNPS
451  VTRYEWKPHG AWEEPSLGVL KIQNVGWDNT TIACAACNSW CSWASPVALN
501  VQYAPRDVRV RKIKPLSEIH SGNSVSLQCD FSSSHPKEVQ FFWEKNGRLL
551  GKESQLNFDS ISPEDAGSYS CWVNNSIGQT ASKAWTLEVL YAPRRLRVSM
601  SPGDQVMEGK SATLTCESDA NPPVSHYTWF DWNNQSLPYH SQKLRLEPVK
651  VQHSGAYWCQ GTNSVGKGRS PLSTLTVYYS PETIGRRVAV GLGSCLAILI
701  LAICGLKLQR RWKR**TQSQQG LQENSSGQSF FVR**NKKVRRA PLSEDPHSLG
751  CYNPMMEDGI SYTTLRFPEM NIPRTGDAES SEMQRPPPDC DDTVTYSALH
801  KRQVGDYENV IPDFPEDEGI HYSELIQFGV GERPQAQENV DYVILKH
```

(i) TQSQQGLQENSSGQSFFVR
(ii) VTPSDAIVR
(iii) QNTFTLNLR
(iv) QAAVTSTSLTIK
(v) IVTCQLQDADGK
(vi) KVTTVIQNPMPIR

B

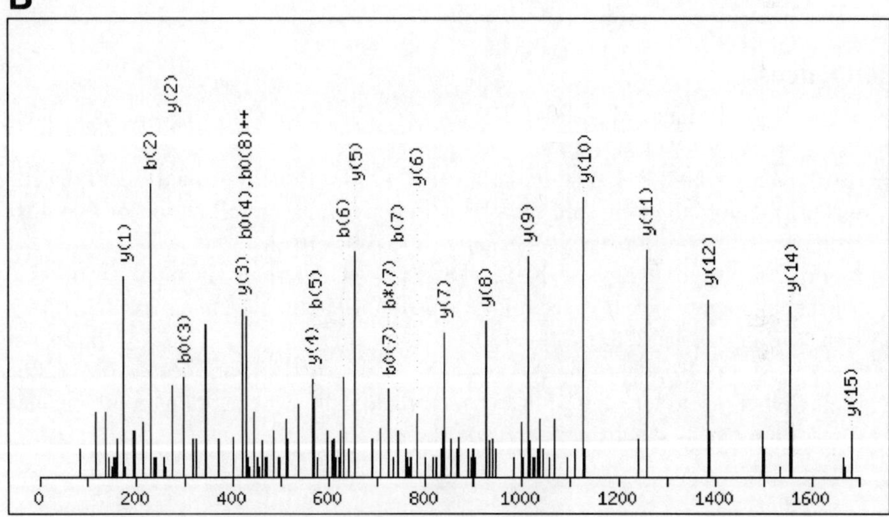

Fig. 4. Analysis of the membrane protein CD22 by proteomics. **(A)** Tandem mass spectrometry (MS/MS) analysis of peptides from plasma membrane fractions identified five peptides from CD22 (i–vi); the peptide sequences are superimposed back on the amino acid sequence of CD22. **(B)** The fragmentation of MS/MS of piptide i. This peptide was doubly charged with a mass-to-charge ratio of 1064.47. The pattern of y ions was used to sequence the peptide.

the yield of total membrane protein; if the yield is less than 10 mg, the SW 60 may be more practical for membrane protein recovery.

11. There are many ways to fractionate a sucrose gradient. We have found that the use of a hooked Pasteur pipet to remove 0.5-mL fractions from the top of the gradient gives reproducible results.

12. If the total protein amount in the solutions being spun down is less than 100 µg, the total yield may be low.

13. It is relatively straightforward to cut 1.5×3 mm slices from a gel lane. A minigel is approx 7 cm, which will result in approx 50 gel slices.

14. The trypsin (Promega, Madison, WI) used in this method has been modified at its lysine residues to generate an active enzyme, which has a greater resistance to autolytic digestion. Trypsin in this kit is provided in 20-µg bottles and must be stored at −20°C. To solubulize the trypsin, add 100 µL of trypsin resuspension buffer and mix well. The trypsin solution can be stored in 12-µL aliquots, which are resuspended in 180 µL of trypsin buffer to give a final amount of 155 ng/well.

15. The acetonitrile is needed to activate the hydrophobic resin.

16. The use of a partial vacuum here is important to ensure the extraction buffer is not filtered too quickly through the hydrophobic resin.

17. This type of method for peptide recovery assumes that 1–2 µg of protein has been subject to proteolysis. If the amount of protein in the gel was less than 100 ng, peptide recovery may be poor.

18. The dried-down peptides can be stored at −20°C.

References

1. Steen, H. and Mann, M. (2004) The ABC's (and XYZ's) of peptide sequencing. *Nat. Rev.* **5,** 699–711.
2. Boyd, R. S., Adam, P. J., and Patel, S., et al. (2003) Proteomic analysis of the cell-surface membrane in chronic lymphocytic leukaemia; identification of two novel proteins, BCNP1 and MIG2B. *Leukemia* **17,** 1605–1612.
3. Adam, P. J., Boyd, R., Tyson, K. L., et al. (2003) Comprehensive analysis of breast cancer cell membranes reveals unique proteins with clinical cancer specific expression. *J. Biol. Chem.* **278,** 6482–6489.
4. Pluskal, M. G., Bogdanova, A., Lopez, M., Gutierrez , S., and Pitt, A. M. (2002) Multiwell in-gel protein digestion and microscale sample preparation for protein identification by mass spectrometry. *Proteomics* **2,** 145–150.

12

Bioinformatic Analysis of Adhesion Proteins

Josephine C. Adams and Juergen Engel

Summary

Proteins that mediate cell–cell and cell–extracellular matrix (ECM) adhesion have been fundamental in the evolution of multicellular animals. Fibrillar collagens, proteoglycans, integrins, and cadherins are present in all animals from sponges to mammals, and many other adhesion proteins have arisen during animal evolution. In general, adhesion proteins are large multidomain molecules and are encoded in larger gene families in vertebrates than in invertebrates. With the increasing availability of completely sequenced genomes representing different points on the animal tree of life, bioinformatics is proving to be a very valuable approach for the analysis of the domain organization and relationships of adhesion proteins, which can direct or enhance experimental tests. Here we describe, with examples from the literature, the major methods for identifying sequence homologies; analyzing domain organization and potential for oligomerization; analyzing sequence relationships by multiple sequence alignments and phylogenetic trees, and assessing adhesion proteins as components of functional pathways and tissue systems through comparative genomics.

Key Words: Extracellular matrix; cell adhesion; collagen; thrombospondin; integrin; cadherin; domain; oligomerization; multiple sequence alignment; phylogeny; genomics.

1. Introduction

Proteins that mediate cell–cell and cell–extracellular matrix adhesion have been fundamental in the evolution of multicellular animals by enabling the assembly of groups of cells and the recognition of self from nonself. In modern vertebrates, adhesion proteins comprise the components of the extracellular matrix (ECM), numerous transmembrane or cell-surface proteins that bind to ECM or to other cells, and associated extracellular proteases and their regulators. The importance of adhesion proteins in animal biology is emphasized by the many human genetic and accquired diseases that result from impairment of cell adhesion processes.

From: *Methods in Molecular Biology, vol. 370:*
Adhesion Protein Protocols, Second Edition
Edited by: A. S. Coutts © Humana Press Inc., Totowa, NJ

Gene mutations or knockouts in mice and flies have demonstrated that many adhesion proteins have essential roles in viability or fertility, or contribute to tissue-specific processes and the robustness of the organism in the face of pathological challenges (for examples, *see* **refs.** *1–3*).

Even simple colonial animals such as sponges (*Porifera*) contain ECM components and cell adhesion receptors that are highly related to those of vertebrates. The fibrillar collagens, proteoglycans, and integrins of sponges are a testament to the ancient origin of many adhesion proteins *(4–7)*. Indeed, sequences encoding cadherin cell–cell adhesion receptors have been identified in unicellular choanoflagellate protozoa, which indicates that certain adhesion proteins arose before the origin of metazoa *(8)*. In contrast, the apicomplexan parasites that form another phylum of eukaryotes and undergo many forms of adhesive interaction with animal host cells have proteins that contain domains typical of animal adhesion proteins but do not share the overall domain organization of the animal proteins *(9)*. The great interest and importance of cell adhesion proteins in animal physiology has resulted in their continuous study over decades by all available methodologies. With regard to the evolution of animal adhesion proteins, in this chapter we focus on bioinformatic analysis as an important new addition to the arsenal.

A vast wealth of information on biological entities is now included in databases around the world. Bioinformatics comprises the process by which these data can be accessed, evaluated, and analyzed by computational tools. As such, it is an indispensable research tool that can direct or enhance the quality of bench experiments. Just as one needs molecular biology in embarking on the study of a new protein, bioinformatics should be brought in at the earliest stage of planning a project to inform the experimental design.

As candidates for bioinformatic analysis, adhesion proteins present some special challenges. ECM components are large, multidomain molecules and are frequently oligomeric. As we have mentioned, individual domains are of ancient origin and can also be present in many other proteins of unrelated function (reviewed in **ref.** *10*). The numbers of ECM components and adhesion receptors encoded per genome have expanded throughout animal evolution, such that vertebrate genomes usually encode larger numbers of related family members than invertebrates. Well-studied examples include the cadherin, integrin, and collagen families of adhesion proteins *(5,7,11)*. Because of these complexities, the identification and comparison of family members from different species and phyla can be challenging; yet bioinformatics facilitates this process.

The second major application of bioinformatics that we will discuss is the comparison of sets of genes in genomes with the aim of exploring their combined function and evolution. The number of fully sequenced genomes has increased dramatically in the last few years, and additional sequencing projects are planned that will improve coverage of the eukaryotic tree of life. These data offer unparalleled opportunities to view the relationships of modern organisms at the level of their genomes and predicted proteomes. Many adhesion molecules and ECM components have been studied mainly with regard to their functions in humans and mice. The privilege of having a broad view of adhesion proteins across the animal kingdom provides us with a new perspective for considering how the components of adhesion systems identified in mammals have come to exist as networks under the action of natural selection and can lead us to deeper insights into their biology and phylogeny. The studies of Hutter et al. and Hynes and Zhao *(12,13)* are early examples of the comparative genomics of adhesion proteins.

The methods discussed in this chapter focus on the analysis of adhesion molecules at protein level. We describe the central features of the various programs and the major considerations for applying these methods to any adhesion protein or domain. All of these methods can be carried out through publicly accessible websites and algorithms (**Table 1**). We also illustrate the methods presented by reference to relevant recent publications in the area of cell adhesion. This chapter does not provide explanations of the mathematical calculations that underlie the various programs. For these, the reader is encouraged to read one of the many excellent bioinformatics textbooks, or consult the program descriptions at relevant websites (**Table 1**). New analytical platforms and programs are continually being set up, and the integration of existing databases is an ongoing process. It is worth keeping up with new developments through the various bioinformatics and genomics journals, many of which are open-access. As a caveat to our enthusiasm for the value of bioinformatics, it should be kept in mind that the application of bioinformatics to proteins at best raises predictions of function. To derive real information on protein function, the computational predictions must be accompanied by experimental tests.

2. Methods

2.1. Searching Sequence Databases

For any protein sequence, a primary question in bioinformatics is to identify sequence homologies. Many steps can then be taken to build on this information, as shown in the flowchart in **Fig. 1**. For a first screen of sequence similarities, by far the fastest and most sensitive algorithm is the basic local alignment tool (BLAST), developed by Altschul et al. *(14)*. Electronic versions of BLAST search routines for proteins (BLASTp), nucleotides (BLASTn), or translated sequences (TBLASTN or TBLASTX) are offered by the National Center for Biotechnology Information (NCBI) of the National Institutes of Health (NIH) at http://www. ncbi.nlm.nih.gov. We recommend that fresh users of BLAST start by reading

Table 1
Resources

Websites
CDD: www.ncbi.nlm.nih.gov/structure/
CDART: www.ncbi.nlm.nih.gov/
COG: www.ncbi.nlm.nih.gov/COG
ELM: www.elm.eu.org
European Bioinformatics Institute: www.ebi.ac.uk
ExPASy: www.expasy.ch.
EMBnet: www.ch.embnet.org/
InterPro: www.ebi.ac.uk/interpro/ (or through ExPASy)
NCBI: www.ncbi.nlm.nih.gov
Phylip: http://evolution.genetics-washington.edu/phylip/software.html
Pfam: www.sanger.ac.uk/software/Pfam
PROSITE: www.ebi.ac.uk/ppsearch (or through ExPASy)
SCOP: http://scop.mrc-lmb.cam.ac.uk/scop
SMART: www.smart.heidelberg.de (or through ExPASy)
SwissProt: www.expasy.org/sprot
UCSC Genome Bioinformatics: http://genome.ucsc.edu
San Diego Supercomputer Biology workbench: http://workbench.sdsc.edu

Books
Claverie, J.-M. and Notredame, C. (2003) *Bioinformatics for Dummies.*
  Wiley Publishing, New York.
Doolittle, R. F. (1987) Of *URFS and ORFS: A Primer on How to Analyze Derived*
  Amino Acid Sequences. University Science Books, Herndon, VA.
Felsenfeld, J. (2003) *Inferring Phylogenies.* Sinauer Associates, Sunderland, MA.
Mount, D. A. (2001) *Bioinformatics: Sequence and Genome Analysis.*
  Cold Spring Harbor Laboratory Press, Cold Spring Harbor, NY.

---

the program selection guide in NCBI, which can be located at the BLAST home page. There are now many versions of the BLAST search tool, and it is important to know the advantages of specialized routines. A not-trivial practical advantage of searching via BLAST is that query sequences can be submitted in almost any format, including space and number interruptions, whereas for most other algorithms much time has to be spent in creating the correct file types. A second advantage of conducting searches at NCBI is that the protein and nucleotide databases are held jointly with Europe and Japan and are updated daily, allowing the widest possible initial screening of the sequence of interest. The user can also define his or her own criteria to search a subset of the database. For example, specific categories of database entries, such as expressed sequence tags, can be selected, or the user can specify a search restricted to a particular organism.

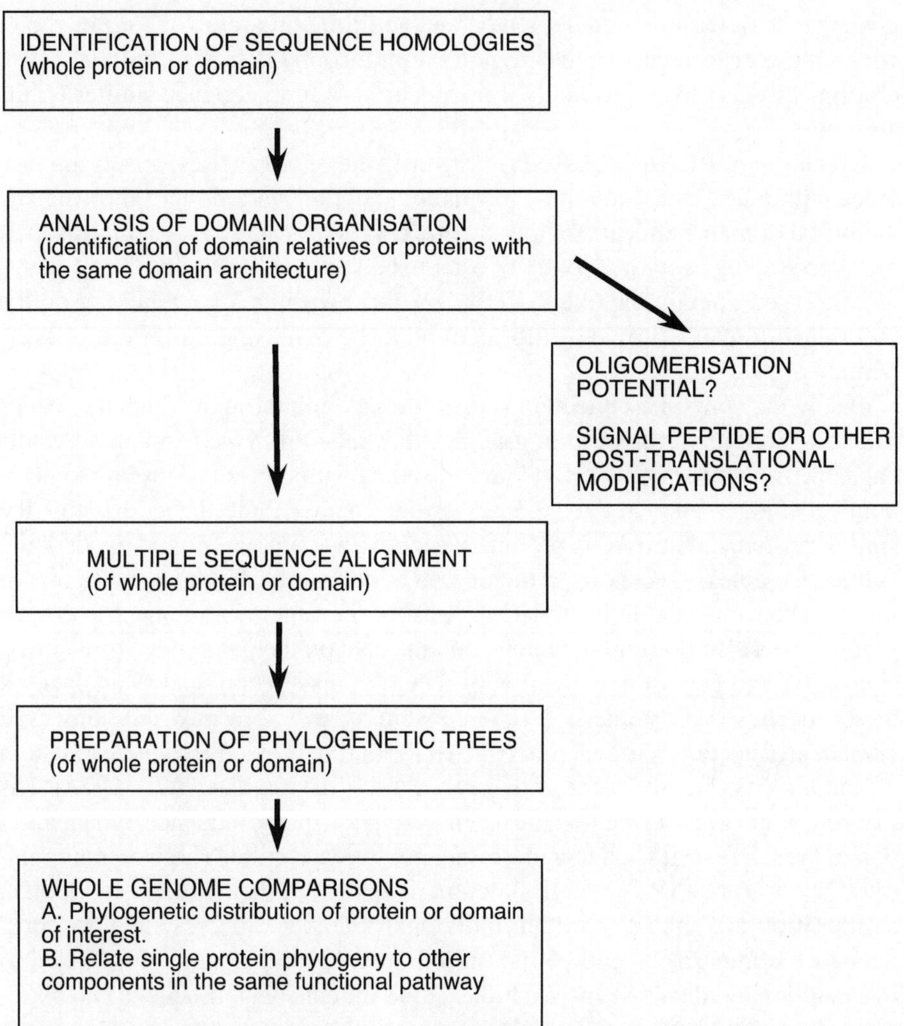

Fig. 1. Relationships between the bioinformatic methods for analysis of adhesion proteins described in this chapter.

The BLAST programs identify regions of similarity between the query sequence and the sequences held within the database(s) searched. The principle of the method is that the BLAST algorithm looks for a region of high similarity with a given word size using a similarity matrix and extends these regions, introducing gaps in the sequence if necessary. The default values provided by the program for the matrix, the word size, the gap punishment, and other criteria should be used in initial searches. It is important to be aware that these values might not apply to the problem under study, and additional searches should be run in which these criteria are varied. A general philosophy is to start with stringent criteria and relax them gradually if no hits or very few sequence similarity hits are found.

Execution of BLAST at NCBI is very fast, and in most cases, results are provided within less than a minute. First, a model of the query protein showing any identified domains appears on the screen. This is produced by a comparison of the query sequence with a library of established sequence patterns for domains, or conserved domain database (CDD; *see* **Subheading 2.2.**). Clicking on the individual domains provides information on the name and definition of these domains.

Below the domain architecture is the "format" link through which the results can be viewed in various text or graphical formats. From the formatted Results page, the link "taxonomy report" provides the results sorted according to taxonomical lineage and organism. Back on the main Results page, a list of the similar proteins identified is provided, each with score values and expectation values (E-values) that indicate the degree of similarity. High scores in bits or low E-values indicate high similarity. A list of the detailed alignment with each protein hit is also presented, which contains data on the percentage of identity, similarity, and gap values. Until very recently, researchers had to analyze all these matches one by one in a time-consuming way, but now the conserved domain architecture retrieval tool (CDART) routine provides a rapid evaluation of homologies directly at the domain level (*see* **Subheading 2.2.**). CDART is reached by clicking on the domain of interest in the query sequence, which takes you to the CDD results page and, from there, selecting the "show domain relatives" icon. In CDART, beautiful comparisons with other proteins by domain composition are shown in which individual domains are presented as icons. Sequence comparisons can be monitored quickly, and each icon is linked to relevant domain database information. Three-dimensional structures can be recalled in cases where such knowledge is available.

The results of a BLAST search will therefore make clear if the sequence under investigation is closely related to any known adhesion proteins. If it is a novel protein, features that would lead one to consider the hypothesis that it functions as an adhesion protein would include: (1) the presence of a domain found in known adhesion proteins, (2) the presence of a signal peptide, and/or (3) evidence of a multidomain architecture. Further corroboration of these indications can be obtained by searching domain databases and conducting additional sequence analyses as described in **Subheading 2.2.**

With regard to searching the genomes of specific organisms, sequence information from many animal genomes is available for BLAST search at NCBI through the "genome biology" link. There are also specific websites for many of the organisms, and because new sequence information or annotation is sometimes available at these sites in advance of release to GenBank, it is worth searching these individually. UCSC Genome Bioinformatics is an excellent site (http://genome.ucsc.edu) where reference sequences and working drafts for a large number of animal genomes can be searched collectively. Results are presented in highly informative formats and can be evaluated with many data analysis tools. Searches of this database proceed by a separate algorithm, BLAT, that is designed to specifically identify close sequence identities (minimum of 95% identity over 40 base pairs for DNA sequences, or 80% identity over 20 amino acids for protein sequences). *See* **Notes 1** and **2** in **Subheading 3.1.**

2.2. Identification and Analysis of Domains

A domain is defined as a self-folding entity that may occur as single or repeated units within a polypeptide sequence *(10)*. The formal definition of a domain requires the recognition of a region within a polypeptide sequence, the demonstration that this region can be efficiently expressed and is functional in isolation, and knowledge of its three-dimensional structure. In practice, the term domain is also applied to regions that have been recognized in multiple proteins on the basis of a characteristic pattern of conserved amino acids. Such regions are also sometimes referred to as modules to emphasize their nature as building units of modular proteins. At one time modules were defined in a stricter sense as domains encoded in a single exon, which could jump between genes by exon–intron shuffling *(15)*. It is now clear that there are not always simple 1:1 relationships between exons and modules, but there are many examples of the conservation of exon structure during the evolution of multidomain proteins *(16)*.

When faced with a novel polypeptide sequence, the identification of domains within the sequence is very helpful in making decisions as to how to express autonomously folding and possibly independently functional portions of the protein or in setting up hypotheses to test for binding partners or functional activities. These issues are particularly relevant to ECM proteins that contain multiple and repeated domains. The list of domains within a protein sequence is referred to as its domain composition. Alternatively, the order of these domains within the polypeptide sequence (referred to as the domain architecture) forms the basis for setting criteria for identifying closely related proteins in the genomes of evolutionarily distant organisms. As an example of the distinction between these terms, the thrombospondins (TSPs) are a gene family of ECM glycoproteins that are highly related because they share a common conserved domain architecture across their C-terminal halves *(17)*. Another domain, the TSP type 1 repeat, is not present in all TSPs but has been widely dispersed in evolution and within the context of a single genome occurs in many proteins that are otherwise unrelated. These proteins constitute a superfamily of domain relatives *(18)*. If the research focus is on the domain itself, it can be informative to study many variants by identifying all other proteins that contain the same domain. An example of such a study is one carried out on the von Willebrand factor A domain *(19)*. By tracking the phylogenetic distribution of this domain and the relationships between proteins containing the domain, the authors charted the evolution of the domain from intracellular to extracellular functions and the expansion of different groups of von Willebrand factor A domain-containing proteins in different phyla. In the alternate situation of identifying a putative novel domain, the comparison of multiple related domains is the basis for deriving a consensus amino acid sequence for the domain (*see* **Subheading 2.4.**).

There are many databases of protein domains, and these have been developed from either sequence or structural data. Sequence-based domain databases include the simple modular architecture retrieval tool (SMART) *(20)*, protein family database of alignments and hidden Markov models (Pfam) *(21)*, also CDD *(22)* and CDART platform at NCBI, both of which are linked to the Pfam and SMART databases *(23)* (**Table 1**). SMART includes a complete classification of domains, their characteristic sequence patterns, three-dimensional structures, and other properties and was developed by Schultz et al. *(20,24)*. The InterPro database integrates domain information from Pfam, SMART, and several other domain and motif pattern databases and is a good place to start *(25)*.

A major database of known protein domain structures is the Structural Classification of Proteins (SCOP; http://scop.mrc-lmcb.cam.ac.uk) *(26,27)*. SCOP is manually curated, and protein structures are ordered hierarchically from domains to folds to display their evolutionary relationships. This is very valuable knowledge to have, yet at present the repertoire of protein stuctures covers only about 3% of the protein sequences in GenBank. Although not designed as a domain database, the Clusters of Orthologous Groups of Proteins database (COG; www.ncbi.nlm.nih.gov/COG) is of interest for tracing evolutionary relationships. COG classifies the proteins from completely sequenced prokaryotic and eukaryotic genomes on the basis of their predicted orthologous relationships—an approach that is helpful in tracking members of gene families between genomes *(28)*.

In addition to analysis at the domain level, additional sequence criteria can be applied to evaluate candidate novel adhesion proteins before beginning experiments. Identification of signal peptides is possible through several algorithms accessible from the Expert Protein Analysis System of the Swiss Institute of Bioinformatics (ExPASy; www.expasy.ch.org). The PROSITE database contains a library of motifs for protein modification or functional sites (currently 1322 patterns; **ref. 29**). However, because these are short motifs, results from

PROSITE searches suffer from overprediction. The ELM database provides another approach to motif searching in which contextual filters are applied with the aim of reducing the frequency of false-positives *(30)*.

See **Notes 1** and **2** in **Subheading 3.2.**

2.2.1. Identification of a Single Class of Domain Within a Protein Sequence:

2.2.1.1. IF THE DOMAIN IS ALREADY KNOWN

BLAST search at NCBI will reveal a CDD match to a part of your sequence, e.g.:

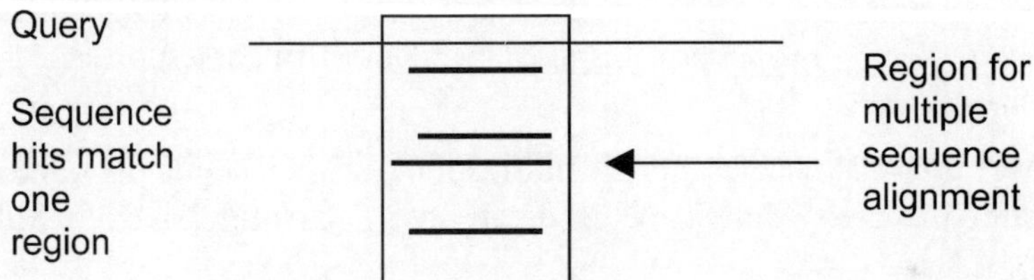

1. CDD provides informative links about the domain and its structure.
2. Go to CDART to identify all other proteins in the NCBI database that contain this domain and compare their domain architectures
3. Search your sequence against the SMART and Inter-Pro domain and motif databases to access other immediate information about the domain. This allows for identification of any other domains/motifs not recognized by CDD criteria.

2.2.1.2. FOR A NOVEL DOMAIN OR TO OBTAIN MORE INSIGHT INTO THE SPECIFIC FEATURES OF A KNOWN DOMAIN WITHIN A SEQUENCE

1. BLAST search at NCBI will reveal a set of sequences that match the same part of your sequence, e.g.:

   (typically, the boundaries of the matching regions will vary somewhat).
2. Extract this portion of sequence from your full-length sequence and from the matching regions identified in other proteins. Label each one as an entry in a FASTA-formatted file.
3. Use this FASTA file to prepare a multiple sequence alignment (MSA) in several programs—CLUSTALW and TCOFFEE are most frequently used (*see* **Subheading 2.4.**).

   Compare the results from CLUSTAL and TCOFFEE to get a view of the robustness of the data.
4. Use the alignments to derive a consensus sequence. The programs allow control of the stringency of the consensus (*see* **Subheading 2.4.**).

5. This consensus sequence can be used to re-search NCBI databases. However, to identify additional examples of the domain, it is more effective to search with distinct individual examples of the domain. PSI-BLAST is an iterative BLAST method designed to help identify domain relatives *(31)*.

6. Experimental follow-up: if the region is related to a known domain, experimental design can be improved based on known structure/function attributes.

If the region appears to be a novel domain: (1) this appreciation will help rational design of domain deletion mutants or functional tests focused on the domain; (2), further experiments can be carried out to establish if the region fulfills all domain criteria (i.e., can it be expressed effectively as an independent unit?; what is its structure?).

In each case a clear evaluation of domain organization and domain characteristics may help distinguish phylogenetics relationships for members of gene families.

2.2.2. Domain Architecture of a Protein Sequence

1. BLAST search and CDD search at NCBI reveals multiple known domains within your sequence, e.g.,

Domain 1   Domain 2   Domain 3

2. These identifications can be confirmed by search of SMART and Inter-Pro, as described above.

3. The order of domains, termed the domain architecture, is an excellent criterion to apply to identify related proteins in NCBI, domain databases, or whole genome databases.

CDART at NCBI enables rapid identification of proteins with the same domain architecture that are within the NCBI databases. (*See* **Notes 3** and **4** in **Subheading 3.2.**)

2.3. Oligomerization Status

The sequence and domain analysis methods described above consider adhesion proteins as single polypeptide chains, i.e., as monomers. An additional level of complexity with many adhesion proteins is that they are secreted as oligomers. Two frequently encountered oligomerization domains are the collagen triple helix and the α-helical coiled-coil (reviewed in **refs.** *32* and *33*). Both of these can be recognized at polypeptide sequence level, providing a basis for predicting nonmonomeric status. The collagen triple helix is readily recognized by eye as repeating triplets of Gly-X-Y, in which X and Y stand for any amino acid and are frequently proline or hydroxyproline. The coiled-coil is based on a characteristic heptad repeat $(abcdefg)_n$ in which the a and d positions are predominantly hydrophobic residues. Other positions are filled by polar residues. By this arrangement, two or more α-helices are aligned and stabilized by hydrophobic interactions. Left-handed superhelices predominate in nature *(34,35)*.

A number of algorithms exist to identify coiled-coil regions within proteins: a widely used program is COILS *(36)*, which can be accessed from the ExPASy site (http://www.expasy.org/). This routine is based on the observed frequencies with which different residues occur in positions a to d of many well-established coiled-coil repeats. Examples of two-, three-, or five-stranded coiled-coils are known in ECM proteins. A complexity that cannot be resolved by current computational methods is the prediction of the number of strands that are oligomerized by a particular coiled-coil domain. Therefore, having recognized a coiled-coil domain, the next step needs to be an experimental test of the oligomerization status of the full-length protein or the coiled-coil region in isolation. However, if many members of a protein family have already been studied, it may be possible to make a prediction by considering other adjacent sequence features as well as the coiled-coil domain. For example, although oligomerization can be mediated solely by noncovalent interactions, in many cases ECM oligomers are also stabilized by disulfide bonds between the subunits. These bonds are typically located as pairs at one or other end of the coiled-coil domain. Candidate cysteine residues can be identified on the basis of their position and high conservation.

As an example, the TSP family of vertebrates contains two subgroups: subgroup A, in which the TSP molecules are assembled as trimers, and subgroup B, in which the TSPs assemble as pentamers *(37,38)*. The proteins in both subgroups contain a coiled-coil domain of five heptad repeats. In trimeric TSP-1, the cysteines necessary for interchain disulfide bonds were identified experimentally to be located amino-terminal to the coiled-coil, whereas in the pentameric TSP-3 or TSP-5 these cysteines are located carboxy-terminal to the coiled-coil *(39–41)*. On the basis of this existing knowledge and MSA of the coiled-coil domains, it was possible to predict that *Drosophila* TSP, identified from the sequenced genome, would oligomerize as a pentamer. This prediction was confirmed by subsequent experiments *(42)*.

Globular domains can also be involved in oligomerization. Examples in ECM proteins include the C-terminal globular domains of fibril-forming collagens and the NC1 domain of basement membrane collagen IV. These domains are structurally unrelated but serve the same function of aligning the three-collagen chains in proper registration for triple-helical assembly *(43)*. The oligomerization activity of these and other globular oligomerization domains cannot be defined from the sequence alone and must be explored experimentally *(32)*. In terms of working with a new protein, a region homologous to a domain of known oligomerization potential would be recognized in the protein sequence by the domain-searching routines outlined in **Subheading 2.2.** This identification would suggest directions for experimental tests to validate the hypothesis that the new protein exists as an oligomer.

So far, we have considered ECM proteins that form homo-oligomers. A further complexity is that certain families of ECM proteins, notably the laminins and collagens, assemble as hetero-oligomers. Similarly, integrin adhesion receptors assemble as heterodimers of α and β subunits. In each case, although a repertoire of subunits may be expressed by a single cell, only certain hetero-oligomer combinations are assembled within the secretory pathway. For example, integrin $\alpha6$ subunit partners with the $\beta1$ or $\beta4$ subunits, but not with $\beta3$ subunit, even though integrins containing $\beta3$ are simultaneously expressed on the cell surface. The heterodimerization properties depend on the extracellular domains of the integrin subunits *(44,45)*. In the large laminin family, distinct combinations of heterotrimers are secreted in different tissues (reviewed in **ref. *46***). The rules of assembly are not understood at present and are thus beyond the scope of bioinformatic analysis.

2.4. Construction of Multiple Sequence Alignments

The MSA is a very powerful method for recognizing patterns of conserved amino acids in two or more related sequences. The basic aim of all MSA computational procedures is to bring the greatest number of identical amino acids into register in the same column of the alignment. Depending on the purpose of the analysis, alignments are made from full-length sequences, individual domains, or peptide motifs. In each case, by aligning many related examples, segments or positions that are highly conserved between all sequences can be distinguished from positions that are weakly conserved or unique to individual sequences. The most highly conserved positions may represent amino acids that are absolutely critical for function (e.g., the active site of a protease) or amino acids that are critical for structure. The latter situation is immediately clear if there is a protein structure to which the sequence can be compared. Regions of uninterrupted hydrophilic residues may correspond to surface-exposed loops that contain binding sites, for example, the RGD motif, which is the recognition site for integrin binding in many ECM glycoproteins *(47)*. These could be excellent candidate regions for mutagenesis. Again, this situation is readily apparent if there is a structure to which to compare the alignment; if not, secondary structure prediction programs can be used to test if the region is predicted to form a surface-exposed loop.

The computation of sequence alignments involves some challenging issues, and many different programs have been devised that address these in different ways. The key issues are that (1) achieving alignment of more than two sequences may necessitate introducing gaps and (2) there must be a set of rules for weighting matching residues vs conservative and nonconservative substitutions and gaps in the overall alignment. The most widely used MSA program, CLUSTALW or CLUSTALX, uses a progressive local alignment method in which the most similar sequences are aligned first after which others are added into the alignment *(48)*. In iterative methods such as DIALIGN, an initial alignment is made and then revised *(49)*. TCOFFEE is a more recently developed progressive local alignment method in which accuracy has been improved by comparing the alignment across the whole set of sequences *(50)*. TCOFFEE can also be used to compare alignments produced by different programs or to combine sequence and structural alignments *(51)*. These programs can be accessed from many sites, but for practical use it is advisable to start from a well-maintained platform, which also contains explanations of the different algorithms. ExPASy (**Table 1**) is a frequently used, high-quality proteomics site.

To prepare a MSA, the sequences of interest are brought together in a FASTA file. This file is then entered into the alignment program. As with BLAST, the user has options to alter the stringency criteria for building the alignment. CLUSTALW and TCOFFEE both provide many output file formats that can be used as the basis for text or graphical presentation of the alignment. The "html" output of TCOFFEE provides an instant color-coded view of alignment quality. For publication, it is convenient to import the aligned CLUSTALW format (TCOFFEE also provides this format) into Boxshade, which allows display of a consensus sequence and user-determined variations for shade or color coding. Boxshade is available at both EMBnet and SDSC Biology Workbench; Biology Workbench can handle larger numbers of sequences (http://workbench.sdsc.edu) (**Table 1**). (*See* **Notes 1–4** in **Subheading 3.4.**)

2.5. Phylogenetic Analysis From Sequence Comparisons

Sequence alignments are the prerequisite for constructing phylogenetic trees from sequence information. A simple tree is already contained in the order in which the sequences appear in the MSA. In the standard aligned output file of CLUSTALW, the sequences appear in the order from the most highly related to more distant sequences. If all sequence changes (distances) are caused by mutations during evolution, this means that the more distant sequences originated from genes that have had more time for mutation than the more similar ones. For example, in analyzing the amino-terminal domains of fibrillar collagens, Bornstein restricted his approach to CLUSTALW and derived information on the evolution of these domains *(52)*. The correctness of phylogenetic information therefore depends critically on the quality of the alignment. Any other sequence differences introduced, for example, by errors in the alignment, are highly disturbing and may give rise to completely wrong information and the construction of inaccurate trees.

Phylogenetic information is usually most interesting for rather distant sequences with only 20–30% overall identity. As discussed in **Subheading 2.4.**, such sequences are difficult to align unambiguously. To solve this conflict, it is recommended to align long sequence regions that contain alternating matching and less well-matching regions. The analysis should then be restricted to regions of unambiguous alignment.

In trees, evolutionary distances are not only used for ordering the sequences but also determine the length of the branches. A schematic demonstration of how a tree can be constructed is provided in **Fig. 2**. Many different methods to create a tree have been devised. A frequently used method is distance based. Here, the overall distance between all pairs of residues in the sequence alignment is first calculated and then used as a measure of phylogenetic distance, as shown in the example. Character-based methods like maximum parsimony or maximum likelihood use the individual substitutions to calculate the ancestral relationship and are more computer-intensive and slower. A minimum of branching points (nodes) is used in constructing the tree. Bayesian inference is a more recent approach in which posterior probabilities of phylogenetic trees are derived to infer phylogeny *(53,54)*. Recent studies tend to use several methods in parallel to obtain a consensus view of the most robustly supported parts of the tree. This approach is exemplified in the study of the phylogeny of ADAMTS proteases by van Meir and colleagues *(55)*.

Excellent introductions on how to make a tree are found in the Tree Tutorial of the HIV Sequence Database (http://hiv-web.lanl.gov/), the Centre for Genomics Research of the Medical Research Council (http://www.hgmp.mrc.ac.uk/GenomeWeb/phylogenetic-anal.html/), and in the book by Joseph Felsenstein (**Table 1**). Felsenstein is a pioneer in the field, and his computer package PHYLIP (http://www.med.nyu.edu/rcr/rcr/phylip/seqdoc.html) is the most frequently used algorithm in which many methods and plotting routines are combined. For many users, more recent PHYLIP-based programs such as that at Institute Pasteur (www.pasteur.fr, linked from the scientific research page) may be easier to apply. A statistical test, the so-called bootstrap analysis, is often applied in a phylogenetic analysis. Basically, the sequences to be compared are randomly disordered, and these meaningless sequences are analyzed for comparison. This procedure does not test the validity of a tree but only assesses the information content contained in the correct sequence as compared with a random sequence. Novel mathematical methods for tree design have been recently proposed with the aim of predicting early events with higher reliability. However, it should not be overlooked that the correct alignment of sequences is much more critical than mathematical refinements.

Interpretation of trees is not an easy task and usually needs intuition and independent information (as discussed in **ref. 56**). Early branch points (sometimes

Very clearly, time cannot be inverted. It is therefore surprising that an evolutionary tree can be constructed entirely from information collected from organisms alive today. This is possible from the sequence differences between their DNA and the proteins encoded by this genetic information. The basic assumption is that the number of sequence changes at a given site is proportional to time for homologous molecules or species. Homologous means that the molecules developed from a common precursor. In the mathematical procedures developed for tree construction, it is further assumed that the tree should have a minimum number of branch points. It should be emphasized that the resulting trees are mathematical models, which describe the data in the framework of these not unambiguous assumptions. Errors in alignment will completely obscure the validity of the analysis.

Several models have been developed and modifications of data treatment have been suggested, but even with these improvements phylogenetic trees remain models and cannot reflect the absolute truth. Frequently, several different trees can be obtained from the same data set. A selection is only possible by external criteria.

For a simple demonstration of how a tree can be constructed, assume that the sequences of three protein regions A, B, and C are known, and it is also certain that they are homologous and properly aligned:

	1	2	3	4	5	6	7	8	9	10	11
A	Gly	Val	Ser	Met	Ala	Gly	Cys	Val	Tyr	Glu	Asp
B	Gly	Val	Ser	Met	Ala	Gly	Cys	Met	Tyr	Asp	Glu
C	Gly	Met	Thr	Met	Gly	Gly	Cys	Ala	Phe	Ser	Ser

In this hypothetical example, the residues in positions 1, 4, 6, and 7 are identical, which suggests their conservation. The number of changes are $\Delta_{AB} = 3$ between proteins A and B and $\Delta_{AC} = 7$ between A and C and $\Delta_{BC} = 7$ between B and C, respectively. The Δ-values are the only experimental data on which the analysis is based. The following tree can be constructed from the data:

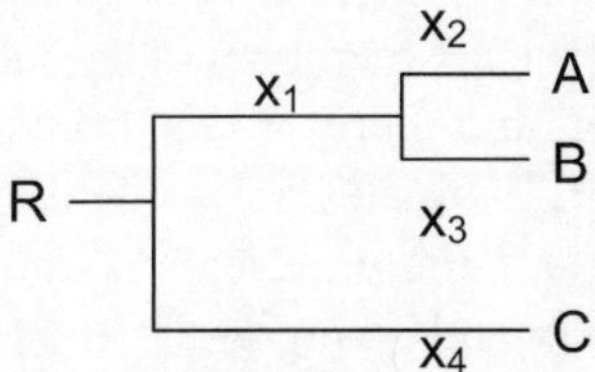

The lengths of the branches x_i (i = 1, 2, 3, 4 in the example) are proportional to time, but the proportionality factor must be determined by independent experiments. The factor may also vary for different proteins and may even be time dependent in itself. The x_i – values are related to the experimental values by:

$$x_2 + x_3 = \Delta_{AB}$$
$$x_1 + x_2 + x_4 = \Delta_{AC}$$
$$x_1 + x_3 + x_4 = \Delta_{BC}$$

Furthermore $x_1 + x_2 = x_1 + x_3 = x_4$

For our numerical example the solution is $x_1 = 2$, $x_2 = x_3 = 1.5$, and $x_4 = 3.5$. R = the root of the tree. *Note*: This example demonstrates only the basic method, and there are other more effective ways to construct a tree (*see* text).

Fig. 2. Construction of phylogenetic trees.

referred to as nodes) are of the most interest, because they may define an early step in which two genes deviated from a common ancestor. On the other hand, early nodes are often ambiguous and are much affected by alignment errors between sequences of low identity. A second problematic issue is the rooting of the tree, which means the definition of the position of the oldest point in the tree *(56)*. The rooting is done either by using independent information, such as dating provided by the fossil record, if available, or by arbitrarily placing the root between the apparently earliest nodes.

If the aim of the tree is to display subgroups within a protein family (i.e., sequence relationships without a time dimension), an unrooted tree may be highly informative and avoid the possible distortion of a root. For example, Nollett et al. classified cadherins from multiple organisms according to the criteria of domain architecture, genomic structure, and phylogenetic comparisions of selected domains. All criteria supported the same groupings, which were then unambiguously displayed by unrooted trees *(11)*.

The relationship between phylogenetic distance and time constitutes the so-called molecular clock. Unfortunately, the pace of this clock may vary for different proteins. For example, epidermal growth factor (EGF) has changed on average by 26% during 100 million years, whereas histone H4 has changed by only 0.1% *(56a)*. Mutation rates may also have been different in different geological eras and under the influence of global geophysical events. In contrast to these facts, the basic assumption of tree construction is a proportionality between divergence in sequence and time. In practice, trees are usually prepared for a single protein family, and it may be assumed that the proportionality factor is approximately the same for all members. The actual value of the factor is usually not known, however, and therefore the branch lengths cannot be translated accurately to real time. Estimates are sometimes obtained by calibrating the tree with an event for which the time scale is known by independent information, for example, by geological dating or the emergence of a species according to the fossil record. For example, in comparing a set of vertebrate TSPs, Lawler et al. calibrated their tree according to the duplication of the *Xenopus laevis* genome *(57)*.

2.6. Adhesion Proteins as Components of Systems Networks

A new area of bioinformatics, comparative genomics, is expanding rapidly as more genomes are sequenced from evolutionarily diverse animals. Currently, 18 completed or shotgun-sequenced metazoan genomes are listed at NCBI, and many more are in progress. Comparative genomics expands the possibilities for bioinformatic and phylogenetic analysis of adhesion proteins from the single molecule to the systems level. It therefore provides a new basis for answering critical questions about the evolutionary development of adhesion proteins and adhesion systems. Such questions cannot be answered from the fossil record and, as discussed in **Subheading 2.5.**, reconstructions from phylogenetic trees are inherently limited in their accuracy by the quality of alignment and by assumptions about the rate of evolution based on the concept of a uniform "molecular clock." Comparative genomics now supplants such heavy reliance on this form of phylogenetic inference by providing accurate data on the predicted proteomes of multiple modern organisms. Large datasets of sequences that are more representative of the animal tree of life can now be compared, and long-held ideas about the evolutionary origin of proteins and pathways can be evaluated by comparing appropriate organisms.

From current knowledge of the context in which a protein functions in one organism (i.e., its synthetic pathway, binding interactions, and functions at cell, tissue, and organism level), its context can be comprehensively evaluated in evolutionarily distant organisms in advance of experimental tests. For example, in addressing whether a protein might have a similar function in a species from another phyla, one would wish to know whether its binding partners are also encoded in the genome of that organism. The collagen family provides a striking example in which biosynthesis and expression of the collagens is only possible with the co-expression of a large repertoire of enzymes and other helpers, the 4-hydroxylases, glycosyltranferases, and N- and C-propeptidases *(58–60)*. Little attention has been paid to comparative phylogenetic analysis of the members of this machinery. The encoding of a collagen chain does not in itself predict collagen expression or function if essential helpers are absent. To address questions such as these, the same BLAST, domain analysis, and phylogenetic techniques as used to study the protein of interest would be applied to identify orthologues of the binding partners or modifying enzymes. To demonstrate the concept of the comparative genomic approach, we discuss two recent examples from the literature where comparative genomics has been applied to functional groups of adhesion proteins.

2.6.1. Evolution of Mineralized Tissue in Vertebrates

Mineralized ECM is crucial for bones, teeth, and body armor (e.g., scales or shells). Despite indications from the fossil record and the widespread occurrence of mineralization in vertebrates and invertebrates, the evolution of mineralized ECM in vertebrates has remained mysterious, in part because many of the ECM components on which mineralization takes place are tissue-specific and have only been identified in mammals. The majority of these proteins, termed the secretory calcium-binding phosphoprotein family, are encoded by a gene cluster that is syntenic in humans, mice, and chickens. On the basis of their exon–intron structures, these genes are thought to have arisen from a common ancestral gene-by-gene duplication *(61)*. Detailed analysis of gene structure and the characteristics of the proteins for homologies with other proteins encoded in the human genome led to the proposal that the ancestral gene was likely a precursor of the 5' region of a gene termed SPARClike 1 (also known as hevin) *(62,63)*. SPARC (secreted protein, acidic and rich in cysteine) is well known as an acidic, calcium-binding protein that is the most abundant non-collagenous ECM component of bone and is also a basement membrane component in *Caenorhabditis elegans* and *Drosophila melanogaster*. SPARClike1 appears to have arisen initially through gene duplication of SPARC and has only been studied in amniotes *(64)*. By tracking the occurance of SPARC and SPARClike1 in the urochordate *Ciona intestinalis* and several fish species, Kawasaki et al. established that the SPARC/SPARClike1 gene duplication most likely occurred within the chordate lineage. The secretory calcium-binding phosphoprotein family appears to have arisen from gene duplications involving domain I of SPARClike1 that occurred after the divergence of cartilaginous and bony fish *(63)*. These data led to the inference that molecular mechanisms for tissue mineralization in mammals arose, in evolutionary terms, quite recently and are distinct from the mechanisms used by invertebrates and cartilagenous fish.

2.6.2. Evolution of the Complement System

Host defense against pathogens is a universal attribute of animals that is provided by the innate immunity system in protostomes and the combined action of innate and adaptive immunity in deuterostomes. Both forms of immunity depend at many levels on the dynamic regulation of proteins that facilitate specific cell-adhesion processes: for example, the recruitment of motile phagocytic cells to sites of injury; engulfment of foreign matter by phagocytosis, and complex cell–cell interactions that activate the adaptive immune response. Central to innate immunity in deuterostomes is the activation of the complement cascade, which serves to recruit inflammatory cells and opsonize and lyse pathogens *(65)*. Because of the complexity of this system and the functional evidence of innate immunity in protostomes, the evolution of the complement cascade has become a question of major interest. Comparisons of fully sequenced genomes, in combination with targeted study of individual components in phylogenetically significant organisms, have identified sequence and functional homologs of the central opsonizing component of the system, C3, and Bf (a serine protease necessary for activating C3), in lampreys, ascidians, and sea urchins, but not in *D. melanogaster* or *C. elegans* *(66)*. The intepretation from these data was that the complement system evolved within the last common ancestor of the deuterostomes. A recent surprise has been the identification of functional and sequence homologs of C3 and Bf in the horseshoe crab, *Limulus polyphemus*, a chelicerate arthropod *(67)*. These data point to a need for further evaluation of the presence of complement components in invertebrates. The new findings also emphasize an important general point about the contributions of bioinformatic analyses to biological understanding: previous conclusions may be very rapidly modified when larger datasets become available. In undertaking any kind of comparative genomics, the largest possible dataset should be assembled. This situation of rapid change is likely to persist for the next few years as additional reference genomes are sequenced. For all the methodologies we have described, we emphasize that results should not be viewed as final. Searches and sequence analyses should be repeated regularly to benefit from the expansion of the sequence and structural databases, the increasing integration between databases, and new computational methods.

Note added in proof. A very powerful application of comparative genomics is the analysis of syntenic relationships between paralogous gene family members in different organisms. This approach is preferable to molecular phylogeny for definitive identification of orthologous members of complex gene families between species. **Ref. *68*** provides a recent example of this approach.

3. Notes

3.1. Searching Sequence Databases

1. In scanning the list of sequence hits from any NCBI search, it is important to be aware that there is a lot of redundancy in the sequence databases. Sequences are deposited either individually by reseachers, as large sets of partial open reading frames from expressed sequence tag sequencing projects, or as hypothetical open-reading frames identified computationally from genomic DNA sequences. Thus, the same predicted protein sequence can be present multiple times in the database in either complete or partial versions. The researcher must sort through all the sequence hits on their own to identify and remove redundant hits. Multiple sequence alignment is an effective and relatively quick way to do this (*see* **Subheading 2.4.**).
2. Although well-studied proteins have extensive literature annotation in GenBank, the automatically generated annotations provided with many GenBank entries should be viewed with caution. Curation of the NCBI databases to reduce redundancy and improve annotation is an ongoing process that proceeds at a slower pace than the growth of the databases. The SwissProt database (www.expasy.org/sprot) is a protein sequence database that is carefully curated to minimize redundancy and includes extensive annotation for each protein in the database. However, it is a much smaller database, (178,940 sequences vs 2,440,549 in GenBank as of April 15, 2005), and so does not allow the same breadth of search as is possible at GenBank. BLASTp searches at NCBI include searches of SwissProt, allowing direction connection to any SwissProt entries returned by the search. For homologies returned by a BLASTn or translated nucleotide database search at NCBI, it is often helpful to repeat the search at SwissProt to quickly obtain detailed information for any of the hits that are included in the SwissProt database.

3.2. Identification and Analysis of Domains

1. Even though there is extensive and growing linkage between the domain databases, the computational algorithms and annotations that underlie each database are distinct. Therefore, a domain, or individual repeats within a set, may be recognized by one program but not another. At present it is always worth searching with the same protein sequence across the different domain databases. For example, the EGF domains of tenascins are not identified by CDD, but are recognized by SMART and InterPro.
2. For sequences that contain a matching region for which a structure has not been determined, a word of warning should be added. The definition of domains from their sequence alone is still subjective and incomplete, and these limitations will affect the results of searches of the domain databases. Matching sequence regions that do not correspond to a domain solved at structural level should be investigated directly at the sequence level, as described in **Subheadings 2.2.1., 2.4.,** and **2.5.**
3. It is important to realize that the annotation of domains into CDART, COG, Pfam, SMART, InterPro, etc. of necessity lags behind the entering of new protein and nucleotide sequences into GenBank or databases of individual genomes. Therefore, to identify the full complement of related proteins in a particular genome, it is more accurate to search GenBank and the genome database directly.
4. Because the sequence-based domain classifications are drawn from GenBank, the protein sequence hits identified will contain the same level of redundancy as GenBank. For example, if 27 hits are identified in SMART or CDART, it cannot be concluded that there are 27 related proteins or that each hit represents a full-length sequence. Analysis of each sequence by BLAST or multiple sequence alignment must be carried out to remove any redundant sequences. The curated dataset can then be used for further analyses of sequence characteristics or phylogeny as described in **Subheadings 2.4.** and **2.5.**

3.3. Construction of Multiple Sequence Alignments

1. The quality of information obtained from a MSA depends very much on the set of input sequences. Correct matches of sequence regions in the MSA will be readily obtained if the sequences are very similar or identical. However, if the sequences compared are more than 90% identical, very little new information will be provided by their alignment. If the sequences are distant (i.e., 25–30% identical), the computational issues of weighting mismatches and gapping will prove limiting to the reliability of the alignment. High-quality alignments are produced from sequences with approx 60–70% overall identity.
2. The choices made in preparing the input dataset will be affected by the purpose for which the alignment is being constructed. For example, to identify candidate regions for mutagenesis, a high-quality alignment of orthologs from within the same phyla might be needed. In contrast, if the alignment is to be used for phylogenetic analysis (*see* **Subheading 2.5.**), the focus might be on distantly related sequences.
3. The number of sequences to be compared is also a consideration. Three or four sequences may not be sufficiently representative, but the accurate alignment of

large numbers of sequences may not be feasible unless they are highly related along their length. As a practical point, most MSA programs place a limit on the number of characters in the entry file; therefore, if sequences are long the number that can be compared is limited. The SDSC Biology Workbench is a web-based toolbox where searches can be combined with analysis and modeling in compatible file formats and large numbers of sequences can be aligned in CLUSTALW.

4. If the full-length sequences are related in only one domain, the input sequences should be cropped to this domain.

Acknowledgments

Research in J.C.A.'s laboratory is supported by the National Institute of General Medical Research, NIH, and the Association for International Cancer Research.

References

1. Sheppard, D. (2000) In vivo functions of integrins: lessons from null mutations in mice. *Matrix Biol.* **19,** 203–209.
2. Bokel, C. and Brown, N. H. (2002) Integrins in development: moving on, responding to, and sticking to the extracellular matrix. *Dev. Cell.* **3,** 311–321.
3. Kakkar, A. K. and Lefer, D. J. (2004) Leukocyte and endothelial adhesion molecule studies in knockout mice. *Curr. Opin. Pharmacol.* **4,** 154–158.
4. Jarchow, J. and Burger, M. M. (1998) Species-specific association of the cell-aggregation molecule mediates recognition in marine sponges. *Cell Adhes. Commun.* **6,** 405–414.
5. Hughes, A. L. (2001) Evolution of the integrin α and β protein families. *J. Mol. Evol.* **52,** 63–72.
6. Exposito, J. Y., Cluzel, C., Garrone, R., and Lethias, C. (2002) Evolution of collagens. *Anat. Rec.* **268,** 302–316.
7. Aouacheria, A., Cluzel, C., Lethias, C., Gony, M., Garrone, R., and Exposito, J. Y. (2004) Invertebrate data predict an early emergence of vertebrate fibrillar collagen clades and an anti-incest model. *J. Biol. Chem.* **279,** 47,711–47,719.
8. King, N., Hittinger, C. T., and Carroll, S. B. (2003) Evolution of key cell signaling and adhesion protein families predates animal origins. *Science* **301,** 361–363.
9. Templeton, T. J., Lyer, L. M., Anantharaman, V., et al. (2004) Comparative analysis of apicomplexa and genomic diversity in eukaryotes. *Genome Res.* **14,** 1686–1695.
10. Ponting, C. P., Schultz, J., Copley, R. R., Andrade, M. A., and Bork, P. (2000) Evolution of domain families. *Adv. Protein Chem.* **54,** 185–244.
11. Nollet, F., Kools, P., and van Roy, F. (2000) Phylogenetic analysis of the cadherin superfamily allows identification of six major subfamilies besides several solitary members. *J. Mol. Biol.* **299,** 551–572.
12. Hutter, H., Vogel, B. E., Plenefisch, J. D., et al. (2000) Conservation and novelty in the evolution of cell adhesion and extracellular matrix genes. *Science* **287,** 989–994.

13. Hynes, R. O. and Zhao, Q. (2000) The evolution of cell adhesion. *J. Cell Biol.* **150,** F89–F96.
14. Altschul, S. F., Gish, W., Miller, W., Myers, E. W., and Lipman, D. J. (1990) Basic local alignment search tool. *J. Mol. Biol.* **215,** 403–410.
15. Patthy, L. (1999) Genome evolution and the evolution of exon-shuffling—a review. *Gene* **238,** 103–114.
16. Liu, M. and Grigoriev, A. (2004) Protein domains correlate strongly with exons in multiple eukaryotic genomes—evidence of exon shuffling? *Trends Genet.* **20,** 399–403.
17. Adams, J. C. (2004) Functions of the conserved thrombospondin carboxy-terminal cassette in cell-extracellular matrix interactions and signaling. *Int. J. Biochem. Cell Biol.* **36,** 1102–1114.
18. Adams J. C. and Tucker, R. P. (2000) The thrombospondin type 1 repeat (TSR) superfamily: diverse proteins with related roles in neuronal development. *Dev. Dynam.* **218,** 280–299.
19. Whittaker, C. A. and Hynes, R. O. (2002) Distribution and evolution of von Willebrand/integrin A domains: widely-dispersed domains with roles in cell adhesion and elsewhere. *Mol. Biol. Cell* **13,** 3369–3387.
20. Schultz, J., Milpetz, F., Bork, P., and Ponting, C.P. (1998) SMART, a simple modular architecture research tool: identification of signaling domains. *Proc. Natl. Acad. Sci. USA* **95,** 5857–5864.
21. Sonnhammer, E. L., Eddy, S. R., and Durbin, R. (1997) Pfam: a comprehensive database of protein domain families based on seed alignments. *Proteins* **28,** 405–420.
22. Marchler-Bauer, A., Anderson, J. B., Cherukuri, P. F., et al. (2005) CDD: a conserved domain database for protein classification. *Nucleic Acids Res.* **33,** D192–D196.
23. Geer, L. Y., Domrachev, M., Lipman, D. J., and Bryant, S. H. (2002) CDART: protein homology by domain architecture. *Genome Res.* **12,** 1619–1623.
24. Letunic, I., Copley, R. R., Schmidt, S., et al. (2004) SMART 4.0: towards genomic data integration. *Nucleic Acids Res.* **32,** D142–D144.
25. Mulder, N. J., Apweiler, R., Attwood, T. K., et al. (2005) InterPro, progress and status in 2005. *Nucleic Acids Res.* **33,** D201–D205.
26. Murzin, A. G., Brenner, S. E., Hubbard, T., and Chothia, C. (1995) SCOP: a structural classification of proteins database for the investigation of sequences and structures. *J. Mol. Biol.* **247,** 536–540.
27. Andreeva, A., Howorth, D., Brenner, S. E., Hubbard, T. J. P., Chothia, C., and Murzin, A. G. (2004) SCOP database in 2004: refinements integrate structure and sequence family data. *Nucleic Acid Res.* **32,** D226–D229.
28. Tatusov, R. L., Fedorova, N. D., Jackson, J. D., et al. (2004) The COG database: an updated version includes eukaryotes. *BMC Bioinform.* **4,** 41.
29. Hulo, N., Sigrist, C. J. A., Le Saux, V., et al. (2004) Recent improvements to the PROSITE database. *Nucleic Acids. Res.* **32,** D134–D137.

30. Puntervoll, P., Linding, R., Gemund, C., et al. (2003) ELM server: A new resource for investigating short functional sites in modular eukaryotic proteins. *Nucleic Acids Res.* **31,** 3625–3630.

31. Altschul, S. F., Madden, T. L., Schaffer, A. A., et al. (1997) Gapped BLAST and PSI-BLAST: a new generation of protein database search programs. *Nucleic Acids Res.* **25,** 3389–3402.

32. Engel, J. (2004) Role of oligomerization domains in thrombospondins and other extracellular matrix proteins. *Int. J. Biochem. Cell Biol.* **36,** 997–1004.

33. Beck, K. and Brodsky, B. (1998) Supercoiled protein motifs: the collagen triple-helix and the alpha-helical coiled coil. *J. Struct. Biol.* **122,** 17–29.

34. Cohen, C. and Parry, D. A. (1994) Alpha-helical coiled coils: more facts and better predictions. *Science* **263,** 488–489.

35. Lupas, A. (1997) Predicting coiled-coil regions in proteins. *Curr. Opin. Struct. Biol.* **7,** 388–393.

36. Lupas, A., Van Dyke, M., and Stock, J. (1991) Predicting coiled coils from protein sequences. *Science* **252,** 1162–1164.

37. Adams, J. C. and Lawler, J. (1993) The thrombospondin family. *Curr. Biol.* **3,** 188–190.

38. Adams, J. C., Tucker, R. P., and Lawler, J. (1995) *The Thrombospondin Gene Family* (Landes, R. G., ed.), Austin, TX, and Springer-Verlag, Berlin.

39. Sottile, J. Seleque, J., and Mosher, D. F. (1991) Synthesis of truncated amino-terminal trimers of thrombospondin. *Biochemistry* **30,** 6556–6562.

40. Efimov, V. P., Lustig, A., and Engel, J. (1994) The thrombospondin-like chains of cartilage oligomeric matrix protein are assembled by a five-stranded α-helical bundle between residues 20 and 83. *FEBS Lett.* **341,** 54–58.

41. Qabar, A., Derick, L., and Lawler, J. (1995) Thrombospondin-3 is a pentameric molecule held together by interchain disulphide linkage involving two cysteine residues. *J. Biol. Chem.* **270,** 12,725–12,729.

42. Adams, J. C., Monk, R., Taylor, A. L., et al. (2003) Characterisation of *Drosophila* thrombospondin defines an early origin of pentameric thrombospodins. *J. Mol. Biol.* **328,** 479–494.

43. Engel, J. and Bächinger, H. P. (2005) Structure, stability and folding of the collagen triple helix. *Topics Curr. Chem.* **247,** in press.

44. Solowska, J., Edelman, J. M., Albelda, S. M., and Buck, C. A. (1991) Cytoplasmic and transmembrane domains of integrin beta 1 and beta 3 subunits are functionally interchangeable. *J. Cell Biol.* **114,** 1079–1088.

45. Frachet, P., Duperray, A., Delachanal, E., and Marguerie, G. (1992) Role of the transmembrane and cytoplasmic domains in the assembly and surface exposure of the platelet integrin GPIIb/IIIa. *Biochemistry* **31,** 2408–2415.

46. Colognato, H. and Yurchenco, P. D. (2000) Form and function: the laminin family of heterotrimers. *Dev. Dyn.* **218,** 213–234.

47. Ruoslahti, E. (1996) RGD and other recognition sequences for integrins. *Ann. Rev. Cell Dev. Biol.* **12,** 697–715.

48. Thompson, J. D., Higgins, D. G., and Gibson, T. J. (1994) CLUSTAL W: improving the sensitivity of progressive multiple sequence alignment through sequence weighting, position-specific gap penalties and weight matrix choice. *Nucleic Acids Res.* **22,** 4673–4680.

49. Morgenstern, B., Dress, A., and Werner, T. (1996) Multiple DNA and protein sequence alignment based on segment-to-segment comparison. *Proc. Natl. Acad. Sci. USA* **93,** 12,098–12,103.

50. Poirot, O., O'Toole, E., and Notredame, C. (2003) Tcoffee: a web server for computing, evaluating and combining multiple sequence alignments. *Nucleic Acids Res.* **31,** 3503–3506.

51. O'Sullivan, O., Suhre, K., Abergel, C., Higgins, D. G., and Notredame, C. (2004) 3DCoffee: combining protein sequences and structures within multiple sequence alignments. *J. Mol. Biol.* **340,** 385–395.

52. Bornstein, P. (2002) The NH(2)-terminal propeptides of fibrillar collagens: highly conserved domains with poorly understood functions. *Matrix Biol.* **21,** 217–226.

53. Yang, Z. and Rannala, B. (1997) Bayesian phylogenetic inference using DNA sequences: a Markov Chain Monte Carlo Method. *Mol. Biol. Evol.* **14,** 717–724.

54. Mitchison, G. J. (1999) A probabilistic treatment of phylogeny and sequence alignment. *J. Mol. Evol.* **49,** 11–22.

55. Nicholson, A. C., Malik, S.-B., Logsdon, J. M., and van Meir, E. G. (2005) Functional evolution of ADAMSTS genes: evidence from analyses of phylogeny and gene organisation. *BMC Evol. Biol.* **5,** 11.

56. Baldauf, S. L. (2003) Phylogeny for the faint of heart: a tutorial. *Trends Genet.* **19,** 345–351.

56a. Atlas of Protein Sequences and Structure, Suppl. 3. (1978) Dayhoff, M. O. ed. National Biomedical Research Foundation.

57. Lawler, J., Duquette, M., Urry, L., McHenry, K., and Smith, T. F. (1993) The evolution of the thrombospondin gene family. *J. Mol. Evol.* **36,** 509–516.

58. Kivirikko, K. I. and Myllyla, R. (1979) Collagen glycosyltransferases. *Int. Rev. Connect. Tissue Res.* **8,** 23–72.

59. Prockop, D. J., Sieron, A. L., and Li, S. W. (1998) Procollagen N-proteinase and procollagen C-proteinase. Two unusual metalloproteinases that are essential for procollagen processing probably have important roles in development and cell signaling. *Matrix Biol.* **16,** 399–408.

60. Myllyharju, J. and Kivirikko, K. I. (2004) Collagens, modifying enzymes and their mutations in humans, flies and worms. *Trends Genet.* **20,** 33–43.

61. Kawasaki, K. and Weiss, K. M. (2003) Mineralized tissue and vertebrate evolution: the secretory calcium-binding phosphoprotein gene cluster. *Proc. Natl. Acad. Sci. USA* **100,** 4060–4065.

62. Delegado, S., Casane, D., Bonnaud, L., Laurin, M., Sire, J.-Y., and Girondot, M. (2001) Molecular evidence for precambrian origin of amelogenin, the major protein of vertebrate enamel. *Mol. Biol. Evol.* **18,** 2146–2153.

63. Kawasaki, K., Suzuki, T., and Weiss, K. M. (2004) Genetic basis for the evolution of vertebrate mineralized tissue. *Proc. Natl. Acad. Sci. USA* **101,** 11,356–11,361.

64. Sullivan, M. M. and Sage, E. H. (2004) Hevin/SC1, a matricellular glycoprotein and potential tumor-suppressor of the SPARC/BM-40/osteonectin family. *Int. J. Biochem. Cell Biol.* **36,** 991–996.
65. Gasque, P. (2004) Complement: a unique innate immune sensor for danger signals. *Mol. Immunol.* **41,** 1089–1098.
66. Nonaka, M. and Yoshizaki, F. (2004) Primitive complement system of invertebrates. *Immunol. Rev.* **198,** 203–215.
67. Zhu, Y., Thangamani, S., Ho, B., and Ding, J. L. (2005) The ancient origin of the complement system. *EMBO J.* **24,** 382–394.
68. McKenzie, P., Chadalavada, S. C., Bohrer, J., and Adams, J. C. (2006) Phylo-genomic analysis of vertebrate thrombospondins reveals fish-specific paralogues, ancestral gene relationships and a tetrapod innovation. *BMC Evol. Biol.* **6,** 33.

13

Analysis of Integrin Dynamics by Fluorescence Recovery After Photobleaching

Bernhard Wehrle-Haller

Summary

Cell migration is a complex cellular behavior that involves the controlled reorganization of the link between the actin cytoskeleton and the extracellular matrix. This mechanical connection is provided by transmembrane receptors of the integrin family. Integrins are heterodimeric receptors that undergo an allosteric switch when activated by external or intracellular signals, providing binding sites for ligands of the extracellular matrix and actin-associated cytoplasmic adapter proteins. Techniques such as fluorescence recovery after photobleaching (FRAP) are used to analyze the remodeling of green fluorescent protein (GFP)-tagged integrin receptors within focal adhesions, demonstrating that the dynamics of the remodeling of integrins in substrate adhesion sites is carefully regulated by extracellular ligands, cytoskeletal adapter proteins, and the actin cytoskeleton. FRAP analysis of GFP-tagged integrins is a tool that allows one to detect and dissect the internal dynamics of apparently immobile focal adhesions, allowing one to perceive the hierarchies and mechanisms of focal adhesion formation and dispersion.

Key Words: Integrins; fluorescence recovery after photobleaching; focal adhesion; focal contact; focal complex; $\alpha v \beta 3$-integrin; green fluorescent protein; integrin dynamics; actin cytoskeleton; B16F1 melanoma; 3T3 fibroblasts; kinetics; model; talin; methods; transfection.

1. Introduction

Fluorescence microscopy is widely used for the qualitative determination of protein distribution in fixed and living cells. Especially the analysis of the cytoskeleton during cell migration and in particular the formation and dispersal of integrin-containing focal adhesion sites has been followed by fluorescence microscopy. The discovery and optimization of the green fluorescent protein (GFP) isolated from the hydrozoa *Aequorea victoria* as a genetically encoded fluorescence reporter has dramatically changed the way to study focal adhesions

From: *Methods in Molecular Biology, vol. 370:*
Adhesion Protein Protocols, Second Edition
Edited by: A. S. Coutts © Humana Press Inc., Totowa, NJ

in living cells. Many cytoplasmic adapter and signaling proteins of focal adhesions have been successfully expressed as GFP fusion proteins and analyzed during the dynamic formation and remodeling of focal adhesions in living cells. In addition, several members of the integrin family of heterodimeric receptors *(1)* have been expressed as GFP-tagged fusion proteins *(2–6)*. The quantitative analysis of the dynamics of these fusion proteins by techniques such as fluorescence recovery after photobleaching (FRAP) is an important tool to further understand the molecular mechanisms that control focal adhesion assembly and disassembly in living cells.

1.1. Focal Adhesion Dynamics

Two different qualitative aspects of focal adhesion protein dynamics can be distinguished. First, the analysis of living cells by fluorescent microscopy can determine the overall formation and dispersal of focal adhesions by following the distribution of large numbers of fluorescent labeled focal adhesion proteins such as the cytoskeletal adapter proteins vinculin and paxillin *(7)*. The second aspect of focal adhesion dynamics happens at a scale that is below the resolution limits of optical microscopes and concerns the dynamic interaction of individual proteins within focal adhesions *(6)*. It is apparent that the behavior and fate of focal adhesions is controlled by the sum of interactions between individual components of focal adhesions. In order to understand and to predict the behavior of focal adhesions and ultimately cell migration, we need to know the dynamic interactions between and the regulatory circuits of individual proteins in the interior of focal adhesions *(8)*.

One technique that allows access to the dynamic properties of proteins within larger biological structures is FRAP. Here we will concentrate on the analysis of the dynamics of integrin molecules at the cell surface and within different types of focal adhesions *(9,10)* using the technique of FRAP.

1.2. Principles of FRAP

FRAP, as well as related photobleaching techniques (inverse FRAP [iFRAP], fluorescent loss in photobleaching [FLIP], and fluorescence localization after photobleaching [FLAP]) *(11–13)*, allows measurement of the movement of fluorescent molecules within or between defined cellular compartments. FRAP takes advantage of the fact that fluorescent molecules such as GFP eventually lose their ability to emit fluorescence when exposed to repeated cycles of excitation and emission (*see* **Note 1**). This effect is referred to as "photobleaching." In FRAP experiments, a small region within a living cell, e.g., a focal adhesion, is photobleached in order to create a spatially separated population of nonfluorescent molecules. In order to measure the mobility of GFP-tagged proteins, images of fluorescent cells are collected over time, during which fluorescent and photobleached molecules redistribute until an equilibrium is reached. By plotting the relationship between fluorescence intensity (F) and time (t), the mobility of the GFP-tagged proteins can be directly measured (**Fig. 1**).

1.3. Parameters and Interpretation of FRAP Experiments

Two descriptive parameters can be deduced from FRAP experiments: first, the "mobile fraction" (M_f) and the related "immobile fraction" (I_f) of fluorescent molecules, and second, the rate of mobility or "halftime of equilibration" ($t_{1/2}$), which depends on the diffusion rate and reversible interactions with binding partners within the analyzed biological structure. The mobile fraction can be determined by comparing the fluorescence intensity in the bleached region after full recovery (F_∞) with the fluorescence intensity before (F_i) and just after bleaching (F_0) (**Fig. 1**). The mobile and immobile fractions are defined as:

$$M_f = (F_\infty - F_0)/(F_i - F_0) \text{ and } I_f = 1 - M_f$$

The mobile fraction can change in response to different circumstances. First, high-affinity binding partners within a multiprotein complex can immobilize a fraction of the bleached protein population, efficiently blocking the exchange with nonbleached fluorescent proteins in such a way that fluorescence recovery appears incomplete. However, in a living cell, after a sufficiently long time of observation any protein–protein interaction should be reversible. Therefore, and often for practical reasons, a reduced mobile fraction (**Fig. 1**) is the consequence of the inability to observe the cell for a sufficiently long time period. If the time of observation is shorter than the actual recovery process, the comparison of the mobile fractions at a given time point will give valuable information about the overall dynamic state of different biological structures or their capacity to self-renew.

The value of $t_{1/2}$ (or fluorescence recovery in FRAP) is determined from the kinetic plot (**Fig. 1**) and represents the time period between the end of photobleaching (t_0) at F_0 and the moment when the fluorescence intensity reaches half of the value at full recovery ($F_{\infty 1/2}$).

$$F_{\infty 1/2} = (F_\infty - F_0)/2$$

The lateral mobility of fluorescent membrane components (such as transmembrane proteins of the integrin family) is expressed by the diffusion coefficient D, which is related to $t_{1/2}$. Axelrod and coworkers have described this relationship for the two-dimensional diffusion of lipids or proteins in a membrane. It is defined as:

$$D = (\omega^2/4\, t_{1/2})\, \gamma$$

where ω represents the radius of a focused laser beam at the e^{-2} intensity and γ is a correction factor for the amount of bleaching *(14)*. This equation has been

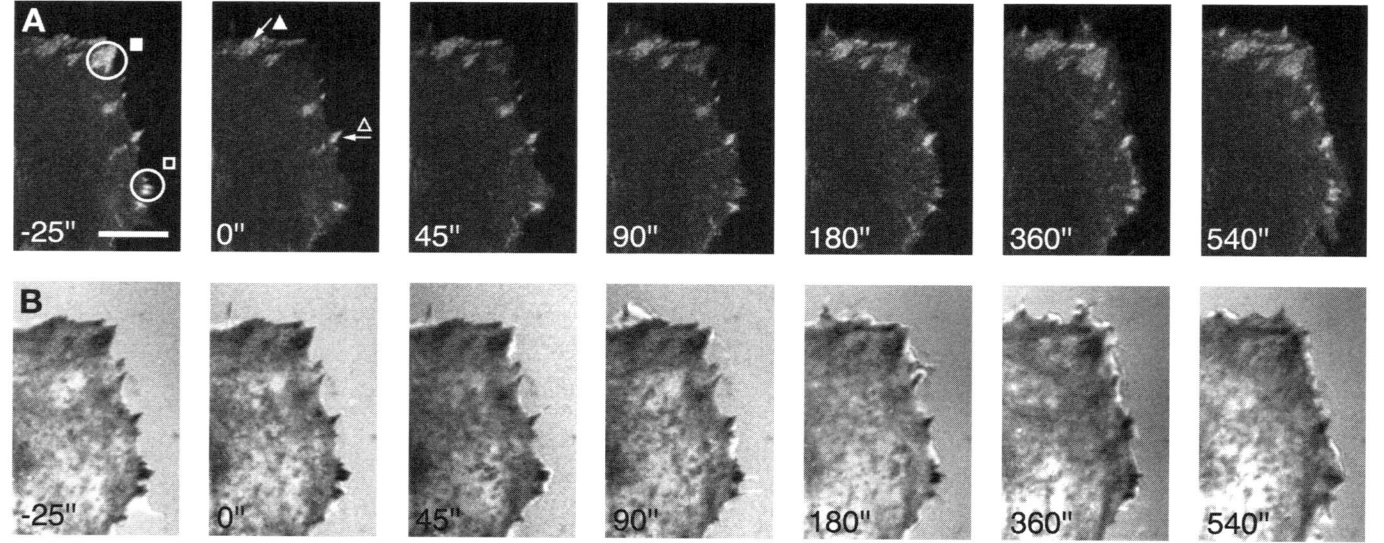

Fig. 1. Fluorescence recovery after photobleaching analysis of green fluorescent protein (GFP)-β3-integrin in focal adhesions of mouse B16F1 melanoma cells. Example of a time series of fluorescence (**A**) and interference reflection (**B**) images of bleached and unbleached peripheral focal adhesions that have been subsequently analyzed for the determination of the fractional fluorescence recovery curve (**F**). Stable GFP-β3-integrin transfected mouse B16F1 melanoma cells were plated overnight and mounted to the heated stage of an inverted confocal laser-scanning microscope (Zeiss, LSM510). Fluorescence (**A**) and IRM (**B**) images were collected simultaneously (*see* **Heading 3.**). The indicated times refer to t_0 (0″) representing the postbleach image. The averaged fluorescence intensities of two bleached focal adhesions (circled in **A** [−25″]) referring to the open and filled squares in (**C**) and control focal adhesions (arrows in **A** [0″]) as well as the background intensity outside of the cell (filled diamonds, F_{bgd}) are traced in (**C**). Note that the measurements of focal adhesion fluorescence intensity were made only in the area covered by the focal adhesion and not over the entire bleached area, excluding pixels that represent membrane or background fluorescence. In (**D**), the fluorescence intensities of both control and bleached adhesions were background subtracted

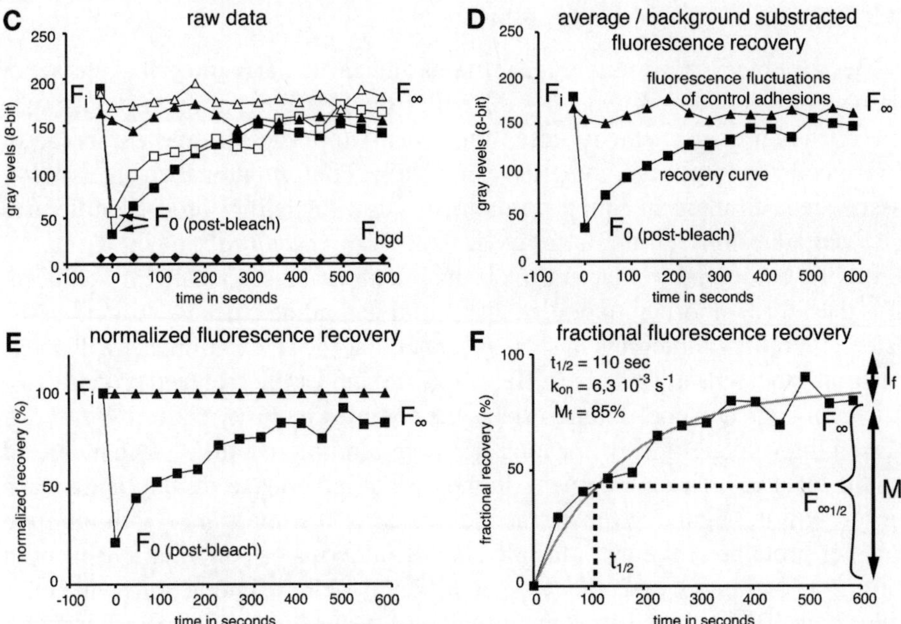

Fig. 1. *(Continued)* and averaged (filled triangles correspond to control and filled squares to the bleached adhesions). In a real experiment, three to five different control and bleached focal adhesions should be analyzed to get a significant average curve. Note that despite the size differences of the analyzed focal adhesions, their recovery curves are similar. In addition, we have noted that in overnight-cultured B16F1 cells even large focal adhesions recover their fluorescence intensity homogeneously. In situations where the pattern and timing of the fluorescence recovery differ notably among focal adhesions, they should be analyzed separately. A different pattern of fluorescence recovery might reveal functional differences between focal adhesions (*see* **Fig. 5**). With the help of the control focal adhesions (filled triangles), the recovery curves were normalized to 100%, compensating thereby for signal fluctuations resulting from focus changes and overall loss of fluorescence intensity in response to observation-dependent bleaching (**E**). The fractional fluorescence recovery curve (**F**) (*see* text) allows extracting the mobile and immobile fractions (M_f, I_f) as well as the half-time of fluorescence recovery ($t_{1/2}$). The latter parameter allows the determination of the rate constant according to a first-order reaction (*see* **Fig. 4**), which results in a theoretical recovery curve (gray line). Bar in **A** = 8 μm.

extremely useful in defining the unrestricted diffusion in a two-dimensional membrane with no recovery from above or below the focal plane. With the help of nonblocking integrin-binding antibodies, Duband and coworkers have measured the diffusion coefficient of the fibronectin receptor in chicken cell membranes to be between 2 and 4×10^{-10} cm^2/s, which is similar to that of other transmembrane proteins and about 100 times lower than that of lipids *(15,16)*.

1.4. What Is Measured With FRAP?

Because many transmembrane proteins and particularly integrins interact or associate with other proteins, it is difficult from FRAP experiments to estimate the diffusion coefficient of "free" integrins within the plasma membrane of living cells. Furthermore, integrins can undergo conformational changes and/or associate with integrin adapter proteins to form the higher ordered structures of focal adhesions. In the latter case, the measured mobility of integrins, for example, compared to that in the plasma membrane, is dramatically reduced. It is therefore important to realize that multiple binding partners will affect the lateral mobility of integrins in such a manner that FRAP experiments will determine an "apparent" diffusion coefficient of the fluorescence-tagged protein. This apparent diffusion coefficient reflects the complex mobility pattern of fast diffusion interrupted by short or long transient binding reactions, which depend on the local cellular environment. In the case of integrins, transient binding can occur with the ligand at the extracellular side of the integrin or with multiple adapter proteins at the cytoplasmic side of integrins or combinations of both (**Fig. 2**) *(10)*. In each case, the respective affinity of these interactions will determine how efficiently the apparent mobility of the integrin is reduced. It is therefore obvious that different interactions of integrins with other proteins will change the lateral mobility of integrins in focal adhesions, complicating the analysis of photobleaching data. However, it also indicates that the measured apparent mobility contains valuable information about the behavior of the integrin and its interaction with binding partners within focal adhesions *(10)*. For example, one can postulate that the strongest binding reaction of integrins with an unknown partner is rate limiting during the FRAP experiment. By modifying individual sites of protein–protein interactions by mutagenesis and subsequent FRAP analysis, it should be possible to determine the rate-limiting binding reactions. In order to understand and predict the behavior of migrating cells, the ultimate goal consists of decrypting the individual affinity constants between binding partners and the rate-limiting interactions that determine the overall behavior of focal adhesions. Based on the obtained kinetic information, mathematical models can be fitted to the experimental data, eventually predicting focal adhesion behavior *(11,17–20)*.

2. Materials

2.1. Cell Culture and Expression Vectors

1. Mammalian expression vector (e.g., pcDNA3, Invitrogen, Basel, Switzerland) for EGFP-tagged α- or β-integrin subunits driven by a suitable promoter (cytomegalovirus , human β-actin promoter) carrying a neomycin antibiotic resistance gene for selection of transfected cells *(6,21,22)*.

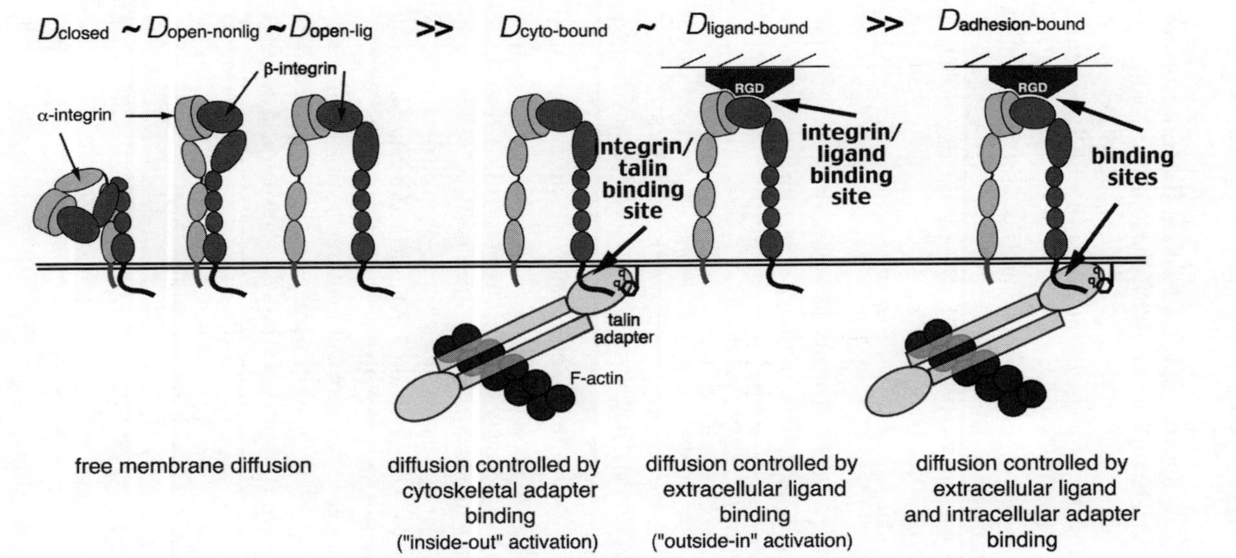

Fig. 2. Altered diffusion coefficients in respect to interaction of integrins with different binding partners. Structural and biochemical data have led to a detailed model of integrin activation and interaction with binding partners at the cytoplasmic tail as well as the extracellular ligand-binding domain *(10)*. Irrespective of the activation state, integrins that are not bound to intracellular adapters or extracellular ligands should exhibit a high diffusion coefficient allowing rapid movement in the plane of the plasma membrane (left-hand side). The apparent diffusion coefficient is reduced, however, as soon as activated integrins begin to transiently interact with intracellular adapters or extracellular ligands, resulting in hop-diffusion-like pattern (center). Alternatively, a reduction in integrin diffusion could be a direct response to the inside-out or outside-in activation and binding of integrins to these binding partners. In any case, once integrins are incorporated into focal adhesions, their apparent diffusion coefficient should be further reduced (right-hand side). One can consider this a reversible binding reaction that occurs independently of the diffusion rate of integrins (*see* **Fig. 4**). (The integrin/ligand/adapter models were built according to information in **refs. *10,45,46*.**)

2. Transfection reagent (e.g., FuGENE 6) (Roche, Basel, Switzerland).
3. Geneticin (G418 sulfate; Invitrogen) dissolved to 100 mg/mL in ddH$_2$O and sterile filtered (*see* **Note 2**).
4. Cell culture medium (Dulbecco's modified Eagle's medium [DMEM]; Invitrogen) supplemented with 10% fetal bovine serum (FBS), glutamine, and antibiotics (Invitrogen).
5. F12 nutrient mixture (Ham) (Invitrogen) supplemented with 10% FBS, glutamine, and antibiotics.
6. Trypsin–ethylene diamine tetraacetic acid (EDTA) in Hank's balanced salt solution (Invitrogen). The concentration of trypsin is 0.05% and that of EDTA 0.53 mM.
7. Cultured cells (e.g., Chinese hamster ovary [CHO], 3T3 mouse fibroblasts, rat embryo fibroblasts [REF52], B16F1 melanoma cells).
8. Fluorescence-activated cell sorter (e.g., Facstar; Becton Dickinson, Basel, Switzerland).
9. Glass-bottom culture wells (e.g., MatTek, Ashland, MA) for culture, transfection, and live imaging of cells expressing fluorescence-tagged integrins.

2.2. Microscope and Software for Image Analysis

1. Inverted laser scanning confocal microscope, e.g., LSM510 on a Zeiss Axiovert 200M (Carl Zeiss Microimaging Inc., Jena, Germany) or Leica TCS SP2 AOBS (Leica Microsystems, Wetzlar, Germany), including software for selection of region-of-interest and bleach routines.
2. High-aperture oil immersion objective (e.g., Zeiss Plan-Neofluar; 63x NA=1,4) (*see* **Note 3**).
3. Heated microscope stage, with CO$_2$ supply and objective heater (*see* **Note 4**).
4. Software to perform basic time-lapse image capture (often provided by the microscope manufacturer). When capturing images with a charge-coupled device (CCD) camera, software such as Openlab (Improvision, UK) or MetaMorph (Molecular Devices) should be used to control image acquisition.
5. Software to analyze the fluorescence intensities of multiple "regions of interest" over time (often included in the image analysis package provided by the microscope manufacturer). Alternatively, software specifically designed to analyze series of fluorescent images can be obtained from several sources (e.g., MetaFluor [Molecular Devices]).
6. Software to perform basic spreadsheet calculations of the obtained fluorescence intensity series (e.g., Excel; Microsoft).

3. Methods

3.1. Integrin Tagging and Expression

3.1.1. Introduction

For a FRAP experiment to provide meaningful data, several basic requirements must be fulfilled by the fluorescence-tagged protein. Most importantly, the fluorescent tag should be applied to a region of the protein in such a way that it does not inhibit its function. In addition, especially important for laterally associating proteins such as integrins, the fluorescent tag is required to remain a monomer and should not contribute to the association reaction. In addition, fluorescent protein-expressing cells should provide images of low noise and a high dynamic range. A large signal-to-noise ratio is required for sensitivity and consistency during data acquisition.

3.1.2. Choosing the Appropriate Integrin Subunit for Fluorescence Tagging

In the mammalian genome, 18 different α-integrin chains and 8 different integrin β subunits have been identified. Excluding alternatively spliced forms of integrins, these subunits form together 24 different integrin heterodimers that are often expressed in redundant and overlapping patterns *(1)*. Because several α- or β-integrin subunits can form heterodimers with multiple β or α chains, respectively, the fluorescent protein tag should be added to a given subunit in such a way as to unequivocally identify one of the multiple species of integrin heterodimers in a cell.

3.1.3. Choosing a Cellular System

Based on the differences in extracellular ligand binding *(10)* or varying regulations of integrin dynamics by the actin cytoskeleton *(5,6)*, FRAP dynamics of each of the 24 different integrin heterodimer is potentially unique. Therefore, it is important to carefully choose the integrin subunit and the cellular model used in order to avoid the simultaneous labeling of different integrin heterodimers. Two distinct expression strategies have been successfully used: the first consists of expressing a new integrin or replacing a missing integrin subunit (via knock-out strategy or clonal selection), while the second approach consists of expressing a GFP-tagged integrin subunit together with the untagged endogenous one. The former system allows expression and analysis of a specific wild-type or mutant fluorescent-tagged integrin in a cellular background that does not express this integrin endogenously, giving at the same time functional information about the biological activity of the GFP-tagged integrin. Such systems include fibroblasts or epithelial cells derived from β1-integrin-deficient embryonic stem cells (GD25 and GE11) *(23,24)*, β3- and β5-deficient CS-1 hamster melanoma cells *(25)*, or CHO cells *(2)*. The latter system has the advantage that even non-functional integrin mutants (e.g., missing the extracellular ligand binding site) can be analyzed in adherent cells *(10)* (*see* **Subheading 3.1.5.**).

3.1.4. Integrin Tagging

Both α and β-integrin subunits have been successfully tagged with fluorescent proteins at the C-termini with spacers of different length *(2,4,6,26)* (*see* **Note 5**).

Because of the complicated extracellular structures of integrins and the allosteric switch of the extracellular integrin domains, N-terminal or intramolecular tagged GFP integrins, although feasible for other type I transmembrane proteins *(27,28)*, have not been reported yet.

3.1.5. Expression of GFP-Tagged $\alpha v \beta 3$ Integrins

As an alternative to the expression of both the α- and β-integrin subunits in CHO cells *(2)*, it is possible to introduce the fluorescent integrin chain within a cellular background that expresses the respective subunit endogenously. In order to fluorescent tag the $\alpha v \beta 3$-integrin heterodimer, we have used cells that already expressed endogenous $\alpha v \beta 3$ integrin (B16F1 melanoma, 3T3 fibroblasts, and REF52) *(5,6,10)*. Upon the introduction of the GFP-tagged mouse $\beta 3$-integrin subunit into these cells, the cell surface levels of fluorescent $\alpha v \beta 3$ integrins are the result of a competition between the GFP-tagged and endogenous $\beta 3$ integrins for a limited amount of αv integrin *(6)* (*see* **Note 6**). Because mouse melanoma and 3T3 fibroblast do not express the αIIb-integrin subunit that could equally pair with the $\beta 3$ subunit, the only fluorescence-labeled integrin is the $\alpha v \beta 3$ heterodimer. In analogy, in order to specifically express a fluorescent form of $\alpha 5 \beta 1$ integrin, it is necessary to introduce a GFP-tagged $\alpha 5$-integrin subunit into the cells of interest *(4,5)* (*see* **Note 7**). Depending on the expression profile of the different α subunits able to pair with the $\beta 1$-integrin subunit, the introduction of GFP-$\beta 1$ integrin would result in the formation of several different species of fluorescent integrins.

3.1.6. Choosing the Appropriate Fluorescent Protein

Integrin heterodimers reversibly associate to form lateral clusters, anchoring the extracellular matrix to the actin cytoskeleton. These clusters are referred to as focal complexes, focal contacts, and focal or fibrillar adhesions *(9)*. Because the lateral clustering of integrins is important for the assembly of focal adhesions, the integrin dynamics can be disturbed when the fluorescent protein tag has the capacity to form dimers or tetramers *(29)*. For these reasons it is preferable to use monomeric fluorescent tags to label integrins (*see* **Note 8**).

The best choice of fluorescent protein for FRAP studies is currently the enhanced green fluorescent protein (EGFP) from Clontech. The monomeric form of EGFP (A206K) *(30)* together with the powerful 488-nm line of Argon lasers offers the largest flexibility between rapid fluorescence inactivation and fluorescence stability for FRAP experiments and prolonged live cell imaging (*see* **Note 9**). For dynamic studies of multiple fluorescent proteins in a single cell, it would be helpful to use different spectral variants of monomeric fluorescent proteins. For FRAP, however, because of the technical limitations of most of the commercially available confocal laser scanning microscopes, the choice of differentially colored fluorescent tags is reduced. In most modern laser scanning confocal microscopes, the intensity of the green lasers is not sufficiently high to induce rapid photobleaching of red fluorescent proteins. However, in combination with EGFP- or yellow fluorescent protein (although less stable than EGFP)-tagged integrin, blue- or red-shifted variants (*see* **Note 8**) *(31)* of fluorescent integrins or other focal adhesion proteins can be used for FLAP experiments *(12,13)* or employed to trace and "calibrate" (*see* **Subheading 3.3.2.**) focal adhesions during FRAP experiments.

3.1.7. Transient vs Stable Expression of Integrins

The generation of cell populations that homogeneously and reproducibly express fluorescent-labeled integrins is critical for performing FRAP experiments. In the case of FRAP analysis of cytoplasmic proteins, transient transfection of the fusion constructs is often sufficient. However, because of the complex maturation process of integrins (*see* **Note 6**) and restricted expression of the mature protein to the membrane compartment of a cell, we prefer to perform FRAP studies of integrins with stably transfected cells *(6,10)*.

3.1.8. Transfection and Selection of Stable Integrin-Expressing Cells

1. Cultured cells with low passage number are grown in DMEM containing 10% FBS, glutamine, and antibiotics. In order to passage cells, they are detached by rinsing with and subsequent incubation at 37°C in trypsin-EDTA solution (the outlined procedure is for B16F1 melanoma cells; other cell types would require some adaptation and optimization of the protocol).
2. When detaching (within 5 min), cells are collected in complete culture medium (DMEM) containing 10% FBS, counted, and seeded at 100,000/well in a six-well culture plate.
3. The following day, the medium of the cells is changed to 1 mL/well of complete medium, while preparing the complex between DNA and transfection reagent.
4. 2.5 µL of transfection reagent (FuGENE 6) is suspended in 100 µL of serum and antibiotic-free culture medium (DMEM). After 5 min, this suspension is added to 1 µg of vector DNA. Mix gently and leave the suspension for 15 min on the bench in order for the plasmid DNA/FuGENE 6 complex to be formed (*see* **Note 10**). This amount is required for one culture well. According to the size and number of culture wells, the amount of DNA and transfection reagent has to be scaled up.
5. To each well already containing 1 mL of complete medium, add 100 µL of the DNA/FuGENE 6 suspension dropwise in a circular motion in order to assure even coverage and return the cells to the incubator.
6. Forty-eight hours later the cells can be detached by trypsin-EDTA and plated into a 10-cm dish containing complete medium supplemented with 1 mg/mL G418 sulfate (optimized for B16F1 cells) (*see* **Note 11**).
7. Replace selection medium at days 4 and 8 of initiation of selection.

8. At day 12 of selection, colonies of transfected cells can be picked individually using cloning rings or cloned by limited dilution into 96-well dishes. Detached cells should be diluted to concentrations of 3 and 10 cells per well (200 µL) and plated in selection medium in 96-well plates. Alternatively, cells can be collected by trypsin–EDTA, blocked and washed in serum-containing medium, and prepared for purification by fluorescence-activated cell sorting (FACS).

9. Prior to FACS purification, cells are washed and blocked in phosphate-buffered saline (PBS) containing 1% bovine serum albumin (BSA) and subsequently incubated on ice for 30 min in PBS/BSA containing diluted antibodies that recognize the extracellular domain of the transfected GFP-tagged integrin subunit (*see* **Note 12**). After washing three times in ice-cold PBS/BSA, a phycoerytrin-coupled secondary antibody is added for 30 min on ice. Following three final washes in ice-cold PBS/BSA, the cells are ready for FACS purification into a tube or dish (batch sorting) or as individual cells into wells of a 96-well plate (*see* **Notes 12** and **13**).

10. Following successful bulk sorting or cloning, cell populations should be expanded and frozen in liquid nitrogen in multiple aliquots (*see* **Note 14**).

Following this protocol, a homogeneous population of bulk sorted GFP-integrin-expressing cells or clones can be obtained within 3–4 wk.

3.1.9. Transient Transfection of GFP-Tagged Integrins

The transient transfection procedure is identical up to **step 5** to that for stable transfected cells (**Subheading 3.1.8.**). However, because fluorescence microscopy or FRAP analysis of integrins is difficult to perform in plastic culture dishes, it is important to subsequently plate transfected cells on glass cover slips. It is therefore important to detach cells with trypsin-EDTA 24 h after transfection and plate them on glass-bottom culture dishes (the glass should have the thickness of a cover slip) in complete DMEM or F12 medium for another 24 h. Alternatively, initial cell plating and subsequent transfection can be directly performed in glass-bottom culture dishes. However, because of different expression levels of the transiently transfected integrins, many of the transfected cells, especially the high expressing cells, cannot be used for FRAP analysis. Typical examples of such cells include those in which the GFP-tagged integrin is retained in intracellular aggregates or within the endoplasmic reticulum.

3.2. Setting Up a FRAP Experiment

3.2.1. Introduction

To study the kinetic behavior of the fluorescence labeled integrins with FRAP, specific regions to be analyzed, such as focal adhesions, need to be selected with a "mask" or "region of interest" tool. This region of interest defines the area that is exposed to a brief but intense excitation pulse to irreversibly inactivate fluorescence emission. Before and immediately after bleaching, the living cell is imaged at low-excitation intensity to determine the initial fluorescence (F_i)

and postbleach intensity (F_0) in the bleached regions (**Fig. 1**). The kinetics of the recovery and redistribution of the fluorescence in the cell is determined from images taken at regular intervals. This series of manipulations can be performed with standard laser scanning confocal microscopes. Alternatively, image capture, before and after photobleaching, can also be performed with a CCD camera. Once collected, images are quantitatively analyzed for changes in fluorescence intensities within the bleached and control regions (e.g., focal adhesions; **Fig. 1**) over time.

3.2.2. Control Experiments

In order to establish the excitation conditions required to completely inactivate fluorescent integrins during a FRAP experiment, it is most convenient to perform this on fixed cells. Cells can be fixed by applying a 4% solution of paraformaldehyde in PBS for 10 min (*see* **Note 15**). After washing in PBS, cells should be mounted onto the confocal microscope in the culture medium that is used for FRAP or time-lapse analysis.

After selection of an appropriate region of interest (e.g., focal adhesion) by using a round or free hand "region of interest" tool, one has to empirically determine the minimum number of iterations (bleach cycles) at maximum laser power required to photobleach the defined region to background levels. In addition, it should be verified that during the period after completed photobleaching, fluorescence does not reappear as a result of reversibility of bleaching or dark states of fluorescent proteins *(11)*. For the analysis of slow recovery processes such as integrin dynamics in focal adhesions, e.g., when using a low frequency of image acquisition, this phenomenon is generally not a problem. These control experiments allow minimized bleaching time and assure that photobleaching is not reversible. A short bleaching time is crucial for FRAP experiments because it limits the redistribution of fluorescent molecules into the bleached region during the bleaching procedure.

Similar control experiments should also be performed with living cells, including the photobleaching of the entire cells. This allows determining whether cell motility and long-term survival is affected by the bleach procedure. In addition, it can be estimated whether *de novo* synthesis of fluorescent molecules can account for an increase in fluorescence during longer FRAP experiments. If this is the case, inhibitors of protein synthesis should be added to the cells at least 1 h prior to the FRAP experiment, taking into account the time difference between GFP protein synthesis and GFP chromophore formation.

3.2.3. Culture Conditions

It is important that cells remain as healthy as possible during the photobleaching experiments. The generation of radicals as a result of laser irradiation especially, but also poor growth conditions or low temperatures, may affect the FRAP experiments (*see* **Note 1**). Optimal conditions for growth and protein mobility include the heating of the microscope stage, medium, and atmosphere as well as the objective to 37°C. In addition, a CO_2 supply to the culture assures a constant pH of the medium (*see* **Note 16**). Prior to mounting the cells onto the inverted microscope, the medium is exchanged to F12 medium supplemented with 10% FBS, glutamine, and antibiotics (*see* **Note 17**). For convenience and easy handling of the cells prior to the FRAP experiment, we use 1-mm-deep culture wells created by gluing a glass cover slip to the bottom of a plastic Petri dish, in which we have previously introduced a circular hole (*see* **Note 18**). This system is convenient because the cells can be transferred without further handling directly from the incubator onto the stage of the inverted microscope.

3.2.4. Image Acquisition and Frame Rate

Once GFP-integrin-expressing cells have been established, FRAP analysis can be undertaken. In comparison to cytoplasmic expressed GFP-fusion proteins, the overall fluorescence of integrin-expressing cells is much weaker, because integrin expression is restricted to the two-dimensional surface of the plasma membranes. In order to produce reliable FRAP data, it is important that the specific signal intensity of the integrins localized within focal contacts is as high as possible compared with the postbleach level (F_0). To reduce noise, a longer pixel dwell time and averaging may be required to get a good image of the integrin fluorescence in focal contacts and membranes. In addition, opening the pinhole increases light collection without affecting the z-resolution of integrins, confined to the two-dimensional cell surface. For FRAP analysis, we have opened the pinhole to two airy units that correspond to 200 µm for a Zeiss LSM510 *(10)* (*see* **Note 19**). Depending on the kinetics of the fluorescence recovery, it is important to find a balance between a high-acquisition frequency, the overall bleaching of the cell, and the signal-to-noise ratio of the acquired images. The latter parameter is determined by the scan rate, averaging, and image resolution and has to be determined empirically. For measuring β3-GFP-integrin recovery in focal contacts, we have opted for relatively long sampling intervals of 30–60 s, a slow scan rate (allowing for averaging [four times] and a pixel dwell time of 1 µs), a moderate image size (512 × 512), and a zoom factor of 1.6–2.0 corresponding to a field width between 70 and 100 µm. Performing conventional time-lapse experiments will allow one to determine the temporal stability of focal adhesions and establish the best imaging conditions.

3.2.5. Duration of the Recovery Period

After bleaching, cellular imaging should be continued until the process of fluorescence recovery reaches a plateau. In B16F1 melanoma cells we found αvβ3-

integrin recovery in focal contacts to be completed after about 10–15 min *(6, 10)*. In other cell types (fibroblasts or endothelial cells), αvβ3-integrin recovery in focal adhesions took longer *(3,5)*. A slower dynamic of the actin cytoskeleton or lower levels of integrin expression may account for the slower integrin recovery. In cells with very large focal adhesions and low integrin expression levels in the plasma membrane, integrin dynamics within focal adhesions should be analyzed by bleaching only a small portion of focal adhesions *(3)* (**Fig. 3**).

The most critical parameter in the analysis of slow recovery processes is to assure a stable focus. This is important because the use of high aperture lenses means that already small focal changes will result in dramatic fluctuations of integrin fluorescence intensities in focal contacts. One way to compensate for intensity fluctuations in response to changes in focus during the recovery period is to normalize the fluorescence intensity in the bleached region to that of reference focal adhesions. A permanent control over the focus at focal adhesions can be obtained by the simultaneous recording of the reflection of the laser light from the medium/cover slip interface through an optimally closed pinhole (1 airy unit). The intensity of the reflection from the glass surface is maximal when the focus is set to the medium/cover slip interface. Moreover, because of the polarized laser light, negative interference between the incoming and the reflected laser light occurring at sites of contact of the cell with the cover slip results in the formation of an interference reflection image typically revealing focal adhesions as dark regions (**Fig. 1B**) *(32,33)*. Using this setting, the focus can always be adjusted to the level of the focal adhesions independently of the fluorescent signal.

In principal, the plateau of the kinetic plot should determine the length of the recovery time. However, sometimes it is more convenient, especially when the FRAP dynamics of integrins are compared under different experimental conditions to maintain acquisition frequency and length of recovery. Using the same FRAP parameters allows one to determine different types of recovery kinetics, for example, caused by integrin mutations *(10)*, or in response to the expression of regulatory proteins of the actin cytoskeleton such as small GTPases of the Rho family *(6)*. Potential differences in the recovery kinetics can then be presented as either a change in the half maximal recovery time (when reaching full recovery) or a change in the mobile fraction determined at a specific time point.

3.3. Data Analysis and Internal Controls

3.3.1. Obtaining Fluorescence Intensity Curves

Once a FRAP image series has been acquired, the fluorescence intensity values must be extracted for bleached as well as unbleached focal adhesions.

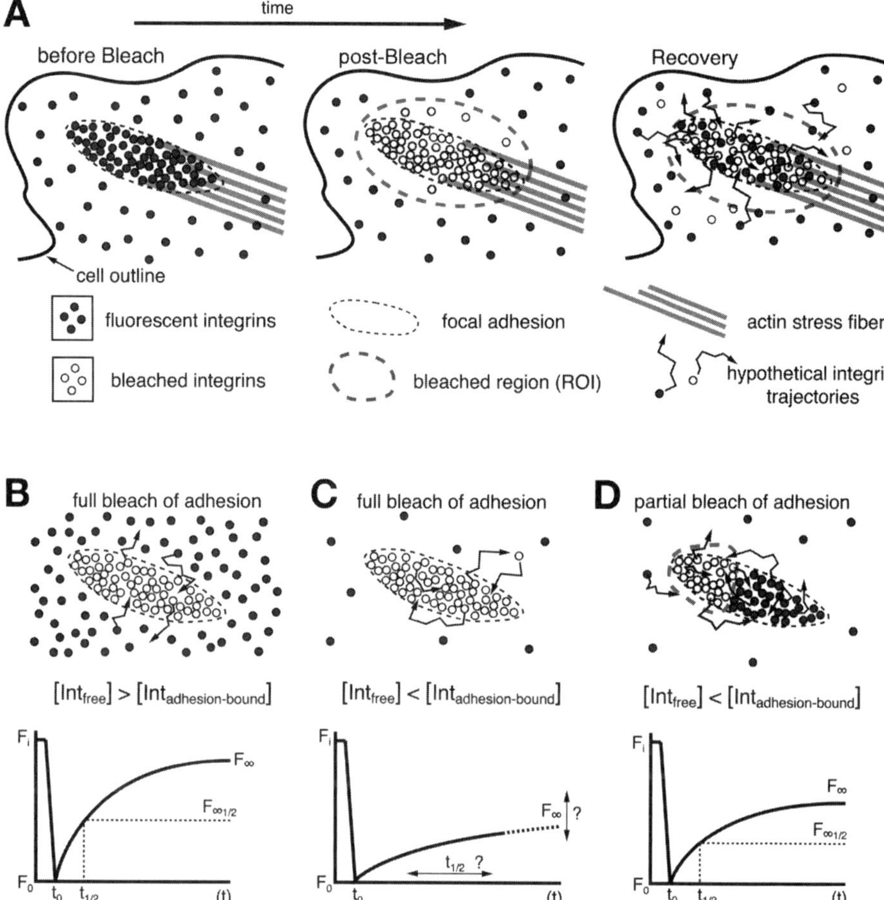

Fig. 3. Schematic view of the fluorescence inactivation and recovery process of green fluorescent protein (GFP)-integrins localized to focal adhesions. (**A**) At the periphery of an imaginary cell, fluorescent integrin molecules (filled circles) are localized in the plasma membrane and in higher concentration within an actin stress-fiber linked focal adhesion (**A**, left graphic) *(6)*. Immediately after bleaching, the nonfluorescent integrins (open circles) are still localized within the focal adhesion (**A**, middle graphic). With time, fluorescent integrins exchange with nonfluorescent ones, inducing a fluorescence recovery in the focal adhesion (**A**, right graphic). Since the diffusion of free integrin molecules outside focal adhesions is much faster than when bound to cytoskeletal adapters inside focal adhesions, the speed of integrin recovery is not influenced by the bleached unbound integrins in the direct vicinity of focal adhesions. Ideally, fluorescence recovery after photobleaching (FRAP) analysis of integrins should be performed under conditions where the concentration of free fluorescent integrins is in great excess over focal adhesion-bound (bleached) integrins (**B**). If the concentration of free integrins in the plasma membrane is reduced, the probability that a fluorescent integrin takes the

Most of the laser scanning confocal microscopes provide software that allows one to select a region of interest (the bleached focal adhesions) in which the averaged fluorescence intensity is determined for each image in the FRAP series. This procedure should be repeated for control focal adhesions as well as for the background fluorescence outside of the cell. Subsequently, the data are imported into Excel, in which further analysis can be done. Alternatively, software to obtain fluorescent intensities over time, for example, from a FRAP image series acquired with a laser scanning microscope or a CCD camera, can be performed with MetaFluor. For technical reasons it is possible for the analyzed cell to shift position during recovery. In this situation it is necessary to register all image frames in respect to a reference point outside of the cell. This can be done using the Openlab software using the Registration module (Improvision, Coventry, UK).

3.3.2. Determination of the Fractional Recovery

In a perfect FRAP experiment the steady-state dynamics of the analyzed cellular structure (e.g., focal adhesion) should not change during the time of the recovery. However, living cells do not hold still, and external or internal signals may modify the steady-state equilibrium of the analyzed adhesive structure. Hence, signal-dependent growth or dispersal of focal adhesions, or a limited pool of fluorescent integrins, would affect either the rate of fluorescence recovery or the immobilized fraction (**Fig. 3**). Thus the presence of focal adhesions that serve as internal controls and that undergo the same influences as the bleached focal adhesions is of critical importance to judge the recovery state and to calibrate each FRAP experiment.

To account for experimental differences, the fractional recovery rate should be determined. The fractional recovery curve is based on a proposition by Axelrod and coworkers *(6,14)* and is easily calculated in Excel. The spreadsheet calculations in Excel allows the handling of large datasets. Typically, from an image series of a FRAP experiment, we determine the averaged ($n = 3$–5) fluorescence intensity of the background outside the cell ($F_{bgd[t]}$) and of bleached

---

Fig. 3. *(Continued)* place of a bleached one is diminished (**C**). In this case, while maintaining a constant integrin exchange rate, nonfluorescent integrins are replaced with nonfluorescent ones. This can give the impression of a slow integrin-recovery process. In such a situation, a partial bleach of a given focal adhesion would nevertheless allow measuring accurate integrin dynamics. Faster fluorescence recovery would be a result of the pool of fluorescent integrin molecules localized to the unbleached region of the focal adhesion (**D**). As a result, the half maximal recovery time can be measured, but at the expense of a reduced mobile fraction. Evidence for such a scenario has been presented by FRAP analysis of focal adhesions in GFP-β3-integrin-transfected endothelial cells *(3)*. ROI, region of interest.

($F_{bleach[t]}$) or unbleached ($F_{ctr[t]}$) focal adhesions or integrin clusters (**Fig. 1**). The unbleached contacts serve as internal controls that allow one to estimate the overall loss of fluorescent molecules from focal contacts as a result of bleaching during the recovery process and to correct for intensity changes in response to changes in focus. Using these internal intensity measurements it is possible to subtract the background fluorescence and to normalize the experimental recovery curve ($R_{norm[t]}$) with the help of the following formula:

$$R_{norm[t]} = (F_{bleach[t]} - F_{bgd[t]})/(F_{ctr[t]} - F_{bgd[t]})$$

The fractional fluorescence recovery curve ($R_{frac[t]}$) further corrects for different postbleach intensities (F_0), which are typically located between 0.05 and 0.25 in the normalized recovery curve. The different postbleach intensities (F_0) are brought to zero, which then permits one to superimpose fractional recovery curves from several different experiments. The fractional fluorescence recovery curve is determined by the following formula:

$$R_{frac[t]} = (R_{norm[t]} - F_0)/(1 - F_0)$$

The half-maximal recovery time as well as the mobile fraction can be directly extracted from this curve (**Fig. 1**).

3.4. Interpretation of Integrin FRAP Data

In order to correctly interpret the obtained FRAP data and to construct a model of integrin dynamics in focal adhesions, it is important to note that the measured recovery rate of the fluorescent integrins depend on two potentially different mechanisms. First, integrins carrying the laser-inactivated fluorophore have to dissociate from the focal contacts, and second, new fluorescent integrins have to be recruited to the preexisting focal adhesions. Thus, the measured FRAP signal is determined by integrin association and dissociation reaction. Normally, in a binding reaction the association (k_{on}) and dissociation (k_{off}) constants are directly linked. However, depending on the potential mechanisms that result in integrin release from (e.g., proteolytic degradation of adapter proteins *[34]*) or recruitment to (e.g., synthesis of PI[4,5]P$_2$ *[10]*) focal adhesions, the apparent association and dissociation constants of integrins may change in function of the fate of focal adhesions (**Fig. 4**).

Although this situation could complicate the interpretation of the FRAP data, it is obvious that the rate of fluorescence recovery is determined by the rate-limiting step in integrin association/dissociation dynamics.

3.4.1. Analysis of Integrin Dissociation Reaction by iFRAP

One technique to directly analyze the dissociation reaction of integrins is to perform an iFRAP experiment. This technique consists of bleaching the pool of fluorescent proteins outside of the analyzed contact. The subsequent loss of fluorescence from the unbleached focal contact directly represents the dissociation reaction *(18)*.

3.4.2. Analysis of Integrin Association Reaction

In contrast to iFRAP, it is impossible to directly analyze the theoretically possible association rate of the fluorescent integrins from FRAP experiments as long as other integrins are occupying the respective positions within focal adhesions. Despite this problem, the association rate of integrins can be determined from the observation of focal adhesion assembly. After bleaching of sliding focal adhesions, we have observed *de novo* recruitment and assembly of fluorescent integrins at the inner edge of sliding focal adhesions *(8,10)*. Interestingly, the rate of the increase in integrin fluorescence intensity at this site is about 2.5 times faster than the integrin exchange in preexisting focal adhesions (45 s/110 s) (**Figs. 1F** and **4**) *(10)*. Surprisingly, this integrin assembly rate equaled the rate of integrin exchange in integrin clusters that were not associated with F-actin *(10)*.

These data suggest that the rate-limiting step for integrin exchange in focal adhesions is controlled by integrin dissociation, which is regulated in part by the state of the actin cytoskeleton *(5,6,8)*.

3.5. Toward a Dynamic Model of Focal Adhesions

In order to draw conclusions and to make predictions on the dynamic behavior of focal adhesions and integrins in particular, the dynamic models that can be formulated based on the integrin FRAP data should include the topology of focal adhesions. Because integrins are transmembrane proteins, their incorporation into focal contacts requires access via a plasma membrane. One possible way to form clusters of integrins in the plasma membrane involves lateral diffusion and association into focal contacts. In this model, a small apparent diffusion coefficient results from multiple interactions that link integrins to other proteins in focal adhesions. One prediction of this model is that integrin recovery in the center of large focal adhesions should be slower than at the periphery (**Fig. 5A**). However, because integrin recovery is relatively homogeneous even in large focal adhesions (*see* **Fig. 1A**), it suggests that integrin diffusion across focal adhesions is rapid, only retarded when engaging in extracellular ligand or intracellular adapter binding. This hypothesis would propose that focal adhesions are built of smaller units that allow rapid diffusion of integrins in between (**Fig. 5B**). If this model is correct, integrin recovery within focal adhesions would not reveal a diffusion process (dependent of the surface of the focal adhesion), but rather a first- or second-order binding reaction (*see* **Fig. 4**).

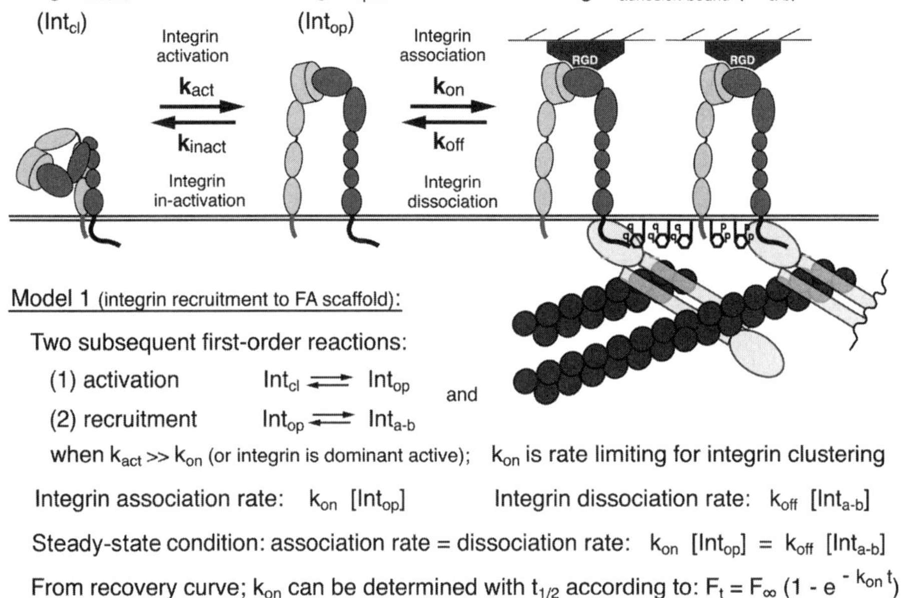

<u>Model 1</u> (integrin recruitment to FA scaffold):

Two subsequent first-order reactions:

   (1) activation $Int_{cl} \rightleftharpoons Int_{op}$ and

   (2) recruitment $Int_{op} \rightleftharpoons Int_{a\text{-}b}$

   when $k_{act} \gg k_{on}$ (or integrin is dominant active); k_{on} is rate limiting for integrin clustering

Integrin association rate: k_{on} [Int_{op}]   Integrin dissociation rate: k_{off} [$Int_{a\text{-}b}$]

Steady-state condition: association rate = dissociation rate: k_{on} [Int_{op}] = k_{off} [$Int_{a\text{-}b}$]

From recovery curve; k_{on} can be determined with $t_{1/2}$ according to: $F_t = F_\infty (1 - e^{-k_{on} t})$

   substituting: $F_{1/2} = \tfrac{1}{2} F_\infty$ results in $k_{on} = \ln 2 / t_{1/2}$

   for integrin recruitment to sliding adhesions ($t_{1/2} = 45s$): $k_{on} = 0{,}015\ s^{-1}$

<u>Model 2</u> (integrin di-/ oligomerization):

Second-order reaction for integrin clustering:

   Focal adhesion assembly: $Int_{op} + Int_{op} \rightleftharpoons Int_{a\text{-}b}$ $K_{eq} = \dfrac{[Int_{a\text{-}b}]}{[Int_{op}]^2}$

   while [Int_{op}] depends on: $Int_{cl} \rightleftharpoons Int_{op}$

Integrin association is controlled by the concentration of activated integrins: [Int_{op}]

   when: [Int_{op}] → 0: *dissociation*   when: [Int_{op}] → ∞: *association*

<u>In general</u>:

   Focal adhesion fate: $k_{on} > k_{off}$ *focal adhesion growth or stabilization*

                             $k_{on} \sim k_{off}$ *focal adhesion is stable*

                             $k_{on} < k_{off}$ *focal adhesion dispersal*

Fig. 4. Dynamic model of integrin clustering and recruitment to focal adhesions. Integrin clustering can be described according to two models, both depending on (1) the activation of integrins and the (2) reversible association with integrin clusters (focal adhesions). In the first model, integrins that are in the inactive or closed conformation diffuse freely in the plasma membrane. After integrin activation (1), integrins are recruited to focal adhesions with a specific association rate constant (k_{on}) and disperse from these

An alternative model of rapid integrin recovery within focal adhesions is based on vesicular transport of integrins *(35–37)*. Potentially, integrins could be directly delivered to or retrieved from focal adhesions by directed exo- and endocytosis. However, although integrin recycling has been observed in several systems, it has not yet been demonstrated that focal adhesions are direct sites of endocytosis or exocytosis of integrins *(38)*.

Irrespective of the mode of integrin transport in the cell, a dynamic model of integrin-dependent focal adhesion formation and turnover will have to include different association and dissociation rates with extracellular ligands and intracellular adapters, the stability of the actin cytoskeleton, as well as different modes of integrin delivery to focal adhesions.

4. Notes

1. Multiple rounds of excitation and emission of fluorescent molecules are associated with the production of radicals. These radicals can locally affect protein and cellular functions and may have deleterious effects on the dynamics of the process to be studied. In contrast to small fluorescent molecules such as fluorescein isothiocyanate, a "protein cage" surrounds the chromophore of the GFP, efficiently capturing radicals generated during fluorescence excitation and emission. Because of this property, the GFP protein is particularly well suited for the noninvasive analysis of dynamic cellular processes by fluorescence microscopy such as FRAP *(39,40)*.
2. Geneticin (G418 sulfate) is obtained as a powder, containing approx 50% active components. Note that the mentioned concentration for the selection of B16F1 melanoma cells is based on the total weight and not of the weight of the active fraction. For each cell type, the concentration of geneticin required to kill cells over a period of a week should be empirically determined. Because of the toxicity

---

Fig. 4. *(Continued)* sites with a dissociation rate constant (k_{off}). Because the fluorescence recovery after photobleaching (FRAP) curves of integrins in peripheral focal adhesions do not change with forced integrin activation *(10)*, we can postulate that the integrin-activation rate is much faster than the integrin-association rate, making integrin association the rate-limiting step in focal adhesion assembly. Assuming that integrin association follows a first-order reaction, we can determine k_{on} from the half maximal integrin association rate in sliding focal adhesions $(t_{1/2} = 45 \text{ s})$ *(10)*. Accordingly, an association rate constant of 6.3 10^{-3} per s can be estimated from FRAP data of peripheral focal adhesions *(see* **Fig. 1**). The second model of integrin clustering is based on the observation that integrin activation results in integrin clustering. Although the mechanism of lateral integrin clustering is complex and not completely understood *(10)*, this mechanism can be formulated according to a second-order reaction. Importantly, the equilibrium of this reaction is dependent on the concentration of activated integrins (Int_{op}). This has the consequence that focal adhesion fates are controlled by the local concentration of activated integrins as well as the respective association or dissociation rate constants.

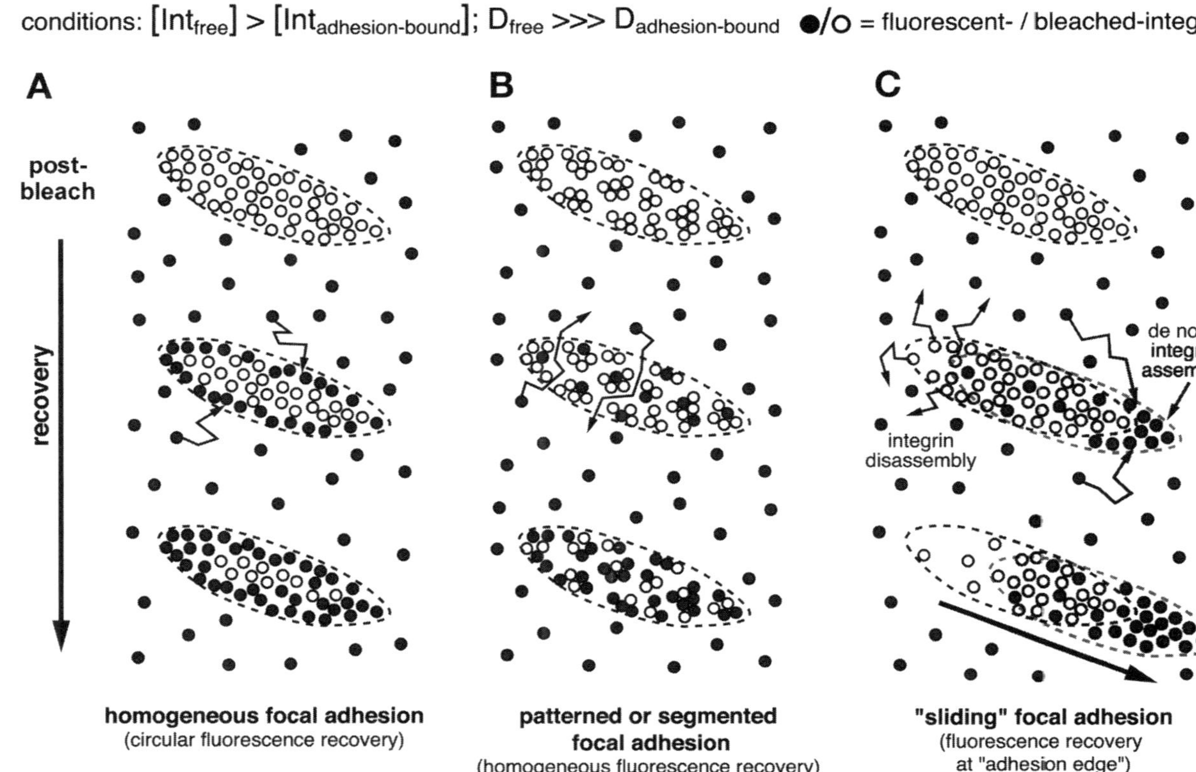

conditions: [Int_free] > [Int_adhesion-bound]; D_free >>> D_adhesion-bound   •/o = fluorescent- / bleached-integrin
A
B
C
post-bleach
recovery
de novo integrin assembly
integrin disassembly
homogeneous focal adhesion
(circular fluorescence recovery)
patterned or segmented focal adhesion
(homogeneous fluorescence recovery)
"sliding" focal adhesion
(fluorescence recovery at "adhesion edge")

of geneticin, the lyophilized content of an entire bottle should be solubilized in ddH$_2$O at once, sterile filtered, and then stored in aliquots at $-20°$C. Alternatively, premade geneticin solutions can be obtained from Invitrogen.

3. The choice of the objective is critically linked to the respective FRAP application. Objectives with a high numerical aperture (e.g., NA = 1.4) are ideal for flat cells or thin biological structures. Because of the good optical resolution along the z-axis, the intensity of the laser beam drops rapidly when outside of the focal plane. For thick objects, this can cause incomplete bleaching outside of the focal plane. Objectives with lower numerical aperture allow a better bleaching along the z-axis, but at the expense of optical resolution.

4. The maintenance of a stable temperature and focus is crucial for FRAP analysis of integrins. Because integrins are confined to the very thin plasma membrane, slight changes in focus (e.g., induced by small changes in objective temperature) will result in a rapid drop of fluorescence intensity. Therefore, a heated chamber that encloses the entire microscope, such as a Ludin Box (Live Imaging Services, Reinach, Switzerland), is preferable to devices that heat the stage, the atmosphere, and the objective independently.

5. It is important that the fluorescent protein tag does not perturb the function of the integrin heterodimer. In one reported example, a human β3-GFP-integrin fusion protein demonstrated partial activation when co-expressed with the nontagged αIIb subunit in CHO cells *(2)*. The reason for this activation is not known, but it might be correlated with the fact that both the α- and the β-integrin subunits were overexpressed. Alternatively, GFP tagging of either one of the C-termini of the two subunits may have destabilized the inactive form of integrins, provoking cytoplasmic tail separations and integrin activation *(41)*.

Fig. 5. *(Opposite page)* The pattern of integrin recovery reveals differences in the structural organization and remodeling of focal adhesions (schematic view of different types of integrin recovery processes in bleached focal adhesions). (**A**) If focal adhesions form a homogeneous structure, integrin molecules will first exchange at the periphery. Subsequently, integrins located at the center of focal adhesions will exchange with a slower kinetic, depending on the presence of fluorescent integrins reaching the periphery of focal adhesions. This type of integrin exchange would result in a circular fluorescence recovery pattern. Interestingly such circular recovery patterns are only rarely observed, suggesting that fluorescent integrins have access to the center of larger focal adhesions (**B**). This observation would predict that focal adhesions are built of smaller units in a patterned or segmented fashion, allowing the diffusion of integrin receptors across these structures (**B**). This "swiss cheese" model of focal adhesions is supported by a recent study demonstrating the formation of stable focal adhesions on nano-patterned substrates that allow integrin binding at a periodicity of 50 nm, leaving ample space for diffusion of unbound integrin receptors *(47)*. (**C**) In the case of sliding focal adhesions, new fluorescent integrins are predominantly recruited to the inner edge of inward sliding focal adhesions, suggesting a polarized assembly and disassembly of integrins in this apparently sliding adhesion sites *(8,10)*.

6. The formation of the integrin heterodimer occurs in the endoplasmic reticulum. Generally, cells express more β-integrin subunits than α subunits, with the consequence that unpaired and chaperone-bound β-integrin subunits accumulate in the endoplasmic reticulum *(42)*. Thus, the surface level of integrins is determined by the amount of α subunits expressed by the cells and a talin-dependent transport of integrin heterodimers to the cell surface *(43)*.

7. Both the ectopic expression of the α- and β-integrin subunits and the introduction of only one GFP-tagged subunit have advantages and disadvantages. While all integrins of the same species are fluorescent labeled in the first case, the second technique allows the expression of wild-type or mutant forms of integrins without dramatically affecting cell adhesion or migration *(10)*. In general, considering the multitude of protein–protein interactions within focal adhesions, it is surprising to see few side effects of GFP-tagging of integrins, making it an acceptable labeling method for studying the dynamics of integrins within focal adhesions.

8. Based on the studies by Zacharias and coworkers, GFP has the tendency to form dimers when restricted to small lipid domains within two-dimensional cell membranes, for example, by adding a GPI-anchor *(30)*. By changing alanine 206 into a lysine (A206K), this GFP–GFP interaction can be suppressed. We expressed both the normal and A206K EGFP-β3-integrin fusion proteins and were not able to see morphological differences or altered integrin clustering in the transfected B16F1 cells (unpublished data). However, it has been demonstrated by fluorescence resonance energy transfer that classical cyan fluorescent protein and yellow fluorescent protein can undergo weak association when tagged to the αL subunit in the αLβ2 integrin, without, however, inducing integrin clustering observable by fluorescence microscopy *(44)*. This suggests that the weak dimerization activity of GFP may only slightly perturb the clustering of GFP-tagged integrins observed during a FRAP experiment. Nevertheless, it is recommended to use monomeric forms of fluorescent proteins when available *(31)*.

9. The suitability of a fluorescent protein for FRAP depends on two parameters. The first parameter is photostability and high quantum yield of the fluorescent protein, which is required for sensitive labeling of proteins and prolonged imaging during the recovery process. The second parameter is the wavelength of the excitation maximum of the fluorescent proteins. The better this excitation maximum overlaps with a high-intensity laser line of the confocal microscope, the faster and more efficient the fluorescent protein can be bleached, gaining temporal resolution for FRAP experiments.

10. For each cell type, the transfection procedure should be adapted for best performance. The purity of the vector DNA is of critical importance. Many commercial plasmid DNA purification kits are available that produce high-quality supercoiled DNA required for efficient transfection.

11. The concentration of G418 sulfate (geneticin) required inducing efficient selection should be determined in control experiments. It is important that cells do not die and detach within 48 h and not later than after 6 d after treatment. This window assures maximal expression of the resistance gene and ample time for cell divisions

that are required for the integration of the vector DNA. Restriction enzyme cleavage of the vector DNA prior to transfection is not required to get integration. Note the incompatibility for the establishment of stable transfected cells with mammalian expression vectors carrying an SV40 origin of replication in cell lines that are immortalized with SV40 large T.

12. During the antibiotic selection process, lasting for about 2 wk, cells that do not express an appropriate amount of GFP-integrins will be eliminated. This is probably linked to the slow proliferation or toxicity of overexpressed transmembrane proteins when accumulating in the endoplasmic reticulum or as insoluble aggregates. Following the initial selection process, a further enrichment for cells that express a maximum of fluorescence-labeled integrins at the cell surface can be achieved by fluorescence-activated cell sorting as follows. GFP-integrin-transfected cells are detached from their culture dish and labeled with an antibody or fluorescence-labeled extracellular ligand specifically binding to the exposed extracellular domain of the GFP-tagged integrins *(6)*. Double-positive cells exhibiting an intermediate level of GFP staining and strong cell surface anti-integrin staining should be selected. Highly GFP-positive cells that express low or no cell surface integrins should be eliminated because they often exhibit bright intracellular pools of fluorescent integrins that hamper the analysis of focal adhesions located at the cell surface. Following this protocol, a homogeneous population of bulk-sorted GFP-integrin-expressing cells can be obtained within 3–4 wk.

13. When sorting individual cells into 96-well plates, take into account the high rate of mortality. In order to assure obtaining stable clones, we generally sort 1, 3, 6, or 10 cells into 24 wells of a 96-well plate.

14. We have noted that in homogeneous GFP-integrin-expressing cell populations, even in the continuous presence of G418 sulfate, the expression levels of GFP-integrin or other GFP-tagged proteins are downregulated. The reason for this is not known but could involve inactivation of the cytomegalovirus promoter by methylation. For practical reasons it is therefore recommended to freeze a large quantity of identical aliquots of a "good" batch of integrin-expressing cells. FRAP and other experiments should then be performed with cells from this original batch and only for a limited amount of passages (e.g., 6–10), corresponding to about a month or two of continuous culturing.

15. A 4% solution of paraformaldehyde in PBS is obtained under continuous stirring and heating to 60°C under a fume hood (attention toxic fumes). After cooling of the solution, aliquots can be frozen at −20°C. Once thawed (briefly warming to 37°C), the paraformaldehyde solution is stable for 1 wk when kept at 4°C.

16. The best choice for heating, stage, objective, and atmosphere consists of heating the whole microscope to 37°C. Commercially available plexiglass chambers (Live Imaging Services, Reinach, Switzerland) offer a comfortable solution, allowing the controlled injection of CO_2 into the circulating atmosphere. One drawback of chambers with large volume is that they cannot be humidified appropriately, which can lead to drying out of the observed cultures. Covering your culture dish with a semi-permeable membrane (we use a wrap obtained from a local supermarket) prevents

evaporation while allowing CO_2 to pass through. Cell adhesion is very sensitive to changes in the pH of the medium. We have found that for fast migrating cells such as B16F1 cells, a pH control by CO_2 is preferable over buffering with 10–20 mM of *N*-2-hydroxyethylpiperazine-*N'*-2-ethanesulfonic acid.

In case the entire microscope cannot be heated, it is important to heat the oil immersion objective. An oil immersion objective is a very effective heat sink when mounted onto a nonheated microscope. Because of the very close contact of the objective to the cells, their effective temperature can easily be 5°C less than that of the heated stage. Objective heaters providing a stable temperature within 1/5 of a degree or better can be very expensive. Considering that small changes in temperature will dramatically affect focus, it is important to dampen the amplitude of a less expensive objective-heating device (e.g., amplitude of 1°C). We indirectly heat the objective through a metal cylinder that surrounds, and is insulated against, the objective.

17. Because of the presence of autofluorescent components in many cell culture media, such as phenol red or riboflavins, background fluorescence may be important, reducing the signal-to-noise ratio of integrin fluorescence. We use F12 medium for time-lapse and FRAP experiments because it exhibits relatively low concentrations of auto fluorescent components when compared with other standard culture media such as DMEM or phenol red-free DMEM.

18. Such culture dishes can be purchased (e.g., MatTek, Ashland, MA). We prepare them from regular 6- or 3.5-cm culture dishes in which we drill a circular hole with a diameter of 12 mm. Precleaned cover slips (cleaning is important to assure a homogeneous coating with extracellular ligands, e.g., vitronectin from serum) (Erie Scientific, Portsmouth, NH) are glued to the bottom of these dishes by dissolving the tissue culture plastic surrounding the hole with about 300 µL of xylene (applied with a Pasteur pipet; use fume hood for manipulation). After about 1 min, the cover slip is pressed into the viscous, partially dissolved plastic and the edges of the cover slip sealed with the same material. After evaporation of the xylene (overnight), the cover slip is cast into the hardened tissue culture plastic. For cell culture, dishes are sterilized by soaking in 70% EtOH (1–2 min) and drying under the tissue culture hood.

19. Although complex bleaching patterns require the use of a laser scanning confocal microscope, it is not necessary to acquire FRAP images in the confocal mode. Thus, image capture before and after the bleaching procedure can also be made in an epifluorescence or total internal reflection mode using a CCD camera coupled to an additional port of the confocal microscope.

References

1. Hynes, R. O. (2002) Integrins: bidirectional, allosteric signaling machines. *Cell* **110,** 673–687.
2. Plancon, S., Morel-Kopp, M. C., Schaffner-Reckinger, E., Chen, P., and Kieffer, N. (2001) Green fluorescent protein (GFP) tagged to the cytoplasmic tail of alphaIIb

or beta3 allows the expression of a fully functional integrin alphaIIb(beta3): effect of beta3GFP on alphaIIb(beta3) ligand binding. *Biochem. J.* **357,** 529–536.

3. Tsuruta, D., Gonzales, M., Hopkinson, S. B., Otey, C., Khuon, S., Goldman, R. D., and Jones, J. C. (2002) Microfilament-dependent movement of the beta3 integrin subunit within focal contacts of endothelial cells. *FASEB J.* **16,** 866–868.

4. Laukaitis, C. M., Webb, D. J., Donais, K., and Horwitz, A. F. (2001) Differential dynamics of alpha 5 integrin, paxillin, and alpha-actinin during formation and disassembly of adhesions in migrating cells. *J. Cell Biol.* **153,** 1427–1440.

5. Hinz, B., Dugina, V., Ballestrem, C., Wehrle-Haller, B., and Chaponnier, C. (2003) Alpha-smooth muscle actin is crucial for focal adhesion maturation in myofibroblasts. *Mol. Biol. Cell* **14,** 2508–2519.

6. Ballestrem, C., Hinz, B., Imhof, B. A., and Wehrle-Haller, B. (2001) Marching at the front and dragging behind: differential alphaVbeta3- integrin turnover regulates focal adhesion behavior. *J. Cell Biol.* **155,** 1319–1332.

7. Webb, D. J., Donais, K., Whitmore, L. A., et al. (2004) FAK-Src signalling through paxillin, ERK and MLCK regulates adhesion disassembly. *Nat. Cell Biol.* **6,** 154–161.

8. Wehrle-Haller, B. and Imhof, B. A. (2002) The inner lives of focal adhesions. *Trends Cell Biol.* **12,** 382–389.

9. Geiger, B., Bershadsky, A., Pankov, R., and Yamada, K. M. (2001) Transmembrane crosstalk between the extracellular matrix—cytoskeleton crosstalk. *Nat. Rev. Mol. Cell Biol.* **2,** 793–805.

10. Cluzel, C., Saltel, F., Paulhe, F., Lussi, J., Imhof, B. A., and Wehrle-Haller, B. (2005) The mechanisms and dynamics of $\alpha v\beta 3$ integrin clustering in living cells. *J. Cell Biol.* **171,** 383–392.

11. Rabut, G. and Ellenberg, J. (2004) Photobleaching techniques to study mobility and molecular dynamics of proteins in live cells: FRAP, iFRAP, and FLIP, in *Live Cell Imaging: A Laboratory Manual*, Vol. 1 (Goldman, R. D. and Spector, D. L., eds.), Cold Spring Harbor Laboratory Press, Cold Spring Harbor, NY, pp. 101–127.

12. Zicha, D., Dobbie, I. M., Holt, M. R., et al. (2003) Rapid actin transport during cell protrusion. *Science* **300,** 142–145.

13. Dunn, G. A., Dobbie, I. M., Monypenny, J., Holt, M. R., and Zicha, D. (2002) Fluorescence localization after photobleaching (FLAP): a new method for studying protein dynamics in living cells. *J. Microsc.* **205,** 109–112.

14. Axelrod, D., Koppel, D. E., Schlessinger, J., Elson, E., and Webb, W. W. (1976) Mobility measurement by analysis of fluorescence photobleaching recovery kinetics. *Biophys. J.* **16,** 1055–1069.

15. Duband, J. L., Nuckolls, G. H., Ishihara, A., et al. (1988) Fibronectin receptor exhibits high lateral mobility in embryonic locomoting cells but is immobile in focal contacts and fibrillar streaks in stationary cells. *J. Cell Biol.* **107,** 1385–1396.

16. Jacobson, K., Ishihara, A., and Inman, R. (1987) Lateral diffusion of proteins in membranes. *Ann. Rev. Physiol.* **49,** 163–175.

17. Dundr, M., Hoffmann-Rohrer, U., Hu, Q., et al. (2002) A kinetic framework for a mammalian RNA polymerase in vivo. *Science* **298,** 1623–1626.

18. Rabut, G., Doye, V., and Ellenberg, J. (2004) Mapping the dynamic organization of the nuclear pore complex inside single living cells. *Nat. Cell Biol.* **6,** 1114–1121.

19. Carrero, G., McDonald, D., Crawford, E., de Vries, G., and Hendzel, M. J. (2003) Using FRAP and mathematical modeling to determine the in vivo kinetics of nuclear proteins. *Methods* **29,** 14–28.

20. Phair, R. D. and Misteli, T. (2001) Kinetic modelling approaches to in vivo imaging. *Nat. Rev. Mol. Cell Biol.* **2,** 898–907.

21. Ballestrem, C., Wehrle-Haller, B., and Imhof, B. A. (1998) Actin dynamics in living mammalian cells. *J. Cell Sci.* **111,** 1649–1658.

22. Ballestrem, C., Wehrle-Haller, B., Hinz, B., and Imhof, B. A. (2000) Actin-dependent lamellipodia formation and microtubule-dependent tail retraction control-directed cell migration. *Mol. Biol. Cell* **11,** 2999–3012.

23. Fässler, R., Pfaff, M., Murphy, J., et al. (1995) Lack of β1 integrin gene in embryonic stem cells affects morphology, adhesion, and migration but not integration into the inner cell mass of blastocytes. *J. Cell Biol.* **128,** 979–988.

24. Gimond, C., van der Flier, A., van Delft, S., et al. (1999) Induction of cell scattering by expression of β1 integrins in β1-deficient epithelial cells require activation of membres of the rho family of GTPases and downregulation of cadherin and catenin function. *J. Cell Biol.* **147,** 1325–1340.

25. Filardo, E. J., Brooks, P. C., Deming, S. L., Damsky, C., and Cheresh, D. A. (1995) Requirement of the NPXY motif in the integrin beta 3 subunit cytoplasmic tail for melanoma cell migration in vitro and in vivo. *J. Cell Biol.* **130,** 441–450.

26. Kim, M., Carman, C. V., and Springer, T. A. (2003) Bidirectional transmembrane signaling by cytoplasmic domain separation in integrins. *Science* **301,** 1720–1725.

27. Paulhe, F., Imhof, B. A., and Wehrle-Haller, B. (2004) A specific endoplasmic reticulum export signal drives transport of stem cell factor (Kitl) to the cell surface. *J. Biol. Chem.* **279,** 55,545–55,555.

28. Wehrle-Haller, B. and Imhof, B. A. (2001) Stem cell factor presentation to c-Kit. Identification of a basolateral targeting domain. *J. Biol. Chem.* **276,** 12,667–12,674.

29. Zhang, J., Campbell, R. E., Ting, A. Y., and Tsien, R. Y. (2002) Creating new fluorescent probes for cell biology. *Nat. Rev. Mol. Cell Biol.* **3,** 906–918.

30. Zacharias, D. A., Violin, J. D., Newton, A. C., and Tsien, R. Y. (2002) Partitioning of lipid-modified monomeric GFPs into membrane microdomains of live cells. *Science* **296,** 913–916.

31. Shaner, N. C., Campbell, I. D., Steinbach, P. A., Giepmans, B. N. G., Palmer, A. E., and Tsien, R. Y. (2004) Improved monomeric red, orange and yellow fluorescent proteins derived from Discosoma sp. red fluorescent protein. *Nat. Biotech.* **22,** 1567–1572.

32. Izzard, C. S. and Lochner, L. R. (1976) Cell-to-substrate contacts in living fibroblasts: an interference reflexion study with an evaluation of the technique. *J. Cell Sci.* **21,** 129–159.

33. Verschueren, H. (1985) Interference reflection microscopy in cell biology: methodology and applications. *J. Cell Sci.* **75,** 279–301.

34. Franco, S. J., Rodgers, M. A., Perrin, B. J., et al. (2004) Calpain-mediated proteolysis of talin regulates adhesion dynamics. *Nat. Cell Biol.* **6,** 977–983.
35. Woods, A. J., White, D. P., Caswell, P. T., and Norman, J. C. (2004) PKD1/PKC micro promotes alphavbeta3 integrin recycling and delivery to nascent focal adhesions. *EMBO J.* **23,** 2531–2543.
36. Roberts, M., Barry, S., Woods, A., van der Sluijs, P., and Norman, J. (2001) PDGF-regulated rab4-dependent recycling of alphavbeta3 integrin from early endosomes is necessary for cell adhesion and spreading. *Curr. Biol.* **11,** 1392–1402.
37. Roberts, M. S., Woods, A. J., Dale, T. C., Van Der Sluijs, P., and Norman, J. C. (2004) Protein kinase B/Akt acts via glycogen synthase kinase 3 to regulate recycling of alpha v beta 3 and alpha 5 beta 1 integrins. *Mol. Cell. Biol.* **24,** 1505–1515.
38. Fabbri, M., Fumagalli, L., Bossi, G., Bianchi, E., Bender, J. R., and Pardi, R. (1999) A tyrosine-based sorting signal in the beta2 integrin cytoplasmic domain mediates its recycling to the plasma membrane and is required for ligand-supported migration. *EMBO J.* **18,** 4915–4925.
39. Chalfie, M. and Kain, S. (1998) *Green Fluorescent Protein: Properties, Applications, and Protocols.* Wiley-Liss, New York.
40. Cole, N. B., Smith, C. L., Sciaky, N., Terasaki, M., Edidin, M., and Lippincott-Schwartz, J. (1996) Diffusional mobility of Golgi proteins in membranes of living cells. *Science* **273,** 797–801.
41. Partridge, A. W., Liu, S., Kim, S., Bowie, J. U., and Ginsberg, M. H. (2005) Transmembrane domain helix packing stabilizes integrin alphaIIbbeta3 in the low affinity state. *J. Biol. Chem.* **280,** 7294–7300.
42. Lenter, M. and Vestweber, D. (1994) The integrin chains beta 1 and alpha 6 associate with the chaperone calnexin prior to integrin assembly. *J. Biol. Chem.* **269,** 12,263–12,268.
43. Martel, V., Vignoud, L., Dupe, S., Frachet, P., Block, M. R., and Albiges-Rizo, C. (2000) Talin controls the exit of the integrin alpha 5 beta 1 from an early compartment of the secretory pathway. *J. Cell Sci.* **113,** 1951–1961.
44. Kim, M., Carman, C. V., Yang, W., Salas, A., and Springer, T. A. (2004) The primacy of affinity over clustering in regulation of adhesiveness of the integrin $\alpha L\beta 2$. *J. Cell Biol.* **167,** 1241–1253.
45. Garcia-Alvarez, B., de Pereda, J. M., Calderwood, D. A., et al. (2003) Structural determinants of integrin recognition by talin. *Mol. Cells* **11,** 49–58.
46. Xiao, T., Takagi, J., Coller, B. S., Wang, J. H., and Springer, T. A. (2004) Structural basis for allostery in integrins and binding to fibrinogen-mimetic therapeutics. *Nature* **432,** 59–67.
47. Arnold, M., Cavalcanti-Adam, E. A., Glass, R., et al. (2004) Activation of integrin function by nanopatterned adhesive interfaces. *Chemphyschem* **5,** 383–388.

14

Double-Hydrogel Substrate as a Model System for Three-Dimensional Cell Culture

Karen A. Beningo and Yu-li Wang

Summary

When cultured on two-dimensional surfaces, most adherent cells show profound differences from those in their native habitats. In addition to chemical factors, it is likely that both physical parameters, such as substrate rigidity, and topographical factors, such as the asymmetry in integrin anchorage, play a major role in the differences. We have designed a simple culture system that provides flexible, adhesive substrates for both dorsal and ventral cell surfaces. Fibroblasts in this system show the spindle or stellate morphology found in native tissues. The ease of preparation, versatility, and optical quality of this model system should greatly facilitate the understanding of cellular behavior and functions in vivo.

Key Words: Cell adhesion; tissue engineering; cell mechanics; hydrogel; polyacrylamide.

1. Introduction

Decades of advances in cell culture have greatly improved the exploration into cellular functions. However, most conventional cell cultures are performed on surfaces of glass or charged polystyrene. Unlike tissues, these materials are stiff and nonporous and impose a striking dorsal–ventral asymmetry in terms of receptor anchorage. Although the simplicity and economy of the approach are directly responsible for the rapid advances in cell biology, it has long been recognized that cells in such cultures show many profound differences from those in vivo. For example, fibroblasts in native tissues are spindle or stellate in shape, with few prominent lamellipodia, stress fibers, or focal adhesions. However, these structures appear within a few hours of plating on a flat polystyrene or glass surface. Many types of cells also lose their differentiated characteristics

From: *Methods in Molecular Biology, vol. 370:*
Adhesion Protein Protocols, Second Edition
Edited by: A. S. Coutts © Humana Press Inc., Totowa, NJ

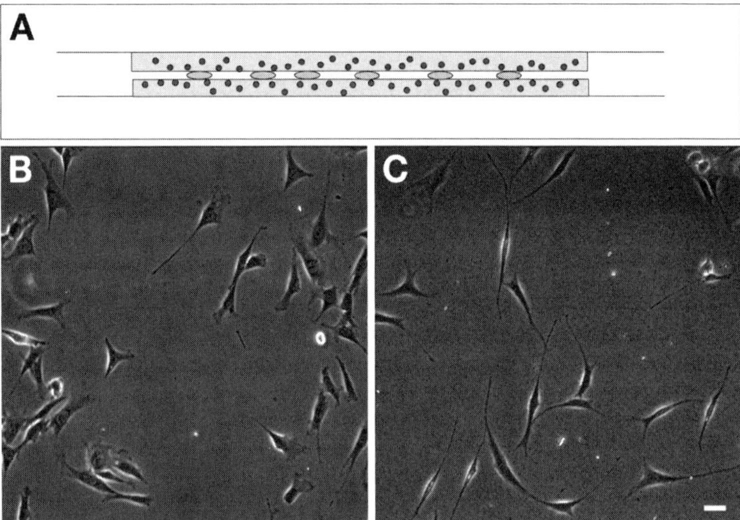

Fig. 1. The double-hydrogel culture system and its effects on fibroblast morphology. (**A**) Cells are cultured between two sheets of extracellular matrix coated polyacrylamide, each about 75 µm in thickness and embedded with 0.5 µm red (top gel) or green (bottom gel) fluorescent microbeads. (**B**) Fibroblasts on fibronectin-coated single hydrogels maintain typical two-dimensional morphology. (**C**) Fibroblasts in the double-hydrogel culture become highly elongated and lack lamellipodia. Bar = 30 µm. (Modified from **ref. 8**.)

and start proliferation. These differences have become an increasing concern to cell biologists as the attention is shifting from the basic growth and motile behavior toward physiological functions.

Over the years a number of model systems have emerged that attempt to address the various shortcomings of the conventional Petri dish system. For example, porous membranes have been used to maintain the polarization of epithelial cells *(1–3)*, whereas three-dimensional (3D) collagen and Matrigel matrices have been developed to promote the native stellate morphology of fibroblasts and acinus morphogenesis of breast epithelial cells *(4–7)*. However, few existing model systems provide optimal conditions for light microscopy, and some of them, such as Millipore or Nucleopore membranes, provide very limited similarities to tissues.

In this chapter we describe a model 3D culture system that possesses several unique properties. Cells are cultured between thin, flexible, porous sheets of hydrogels coated with defined matrix proteins. The system has an optical quality comparable to that of two-dimensional cultures for imaging such structures as the cytoskeleton and adhesion complexes in live cells (**Fig. 1A**). In addition, the system allows easy control of parameters including the type and concentration of extracellular matrix proteins and mechanical properties such as compliance. Although the method was originally designed to address morphological differences between cells in two-dimensional cultures and in native tissues or artificial matrices *(8)*, it may find wide applications in other areas such as gene expression, cancer invasion, and tissue engineering.

2. Materials

2.1. Activated Cover Slips

1. No.1 glass cover slips of suitable dimension for culture chamber (*see* **Note 1**).
2. 22-mm Round cover slips.
3. 0.1 *N* NaOH.
4. 3-Aminopropyltrimethoxysilane (Sigma-Aldrich, St. Louis, MO) stored at 4°C.
5. 0.5% Gluteraldehyde in phosphate-buffered saline (PBS). Gluteraldehyde stocks (70%, Polysciences, Warrington, PA) are stored at 4°C and diluted in PBS immediately before use.
6. Bunson burner.
7. Forceps.
8. Pasteur pipets.

2.2. Polyacrylamide Hydrogels

1. Stock solution of acrylamide (40% w/v, Bio-Rad, Hercules, CA) stored at 4°C.
2. Stock solution of *N,N'*-methylene bisacrylamide (2% w/v, Bio-Rad) stored at 4°C.
3. Stock of 1 *M N*-2-hydroxyethylpiperazine-*N'*-2-ethanesulfonic acid (HEPES), pH 8.5, filtered and stored at 4°C
4. *N,N,N,N'*-Tetramethylethylenediamine (TEMED; Bio-Rad) stored at 4°C.
5. 10% Ammonium persulfate (Bio-Rad), stored desiccated at room temperature.
6. Fluorescent latex beads, 0.5 or 0.2 µm in diameter (Fluospheres, Molecular Probes) stored in the dark at 4°C (*see* **Note 2**).
7. Two pairs of fine-tipped forceps.

2.3. Conjugation of Proteins to the Hydrogels

1. Sulfosuccinimidyl-6-(4'-azido-2'-nitrophenylamino) hexanoate, also known as sulfo-SANPAH (Pierce Chemical, Rockford, IL) (*see* **Note 3**). This product is light sensitive and should be stored desiccated at −20°C and dissolved immediately before use.
2. Dimethylsulfoxide (DMSO) stored desiccated at room temperature.
3. 50 m*M* HEPES, pH 8.5, solution should be filtered and stored at 4°C and warmed to room temperature before use.
4. UV light source with 302-nm bulbs (VWR no. 21476-0101). Use appropriate protection because UV light can burn skin and damage eyes.
5. The protein of interest, typically fibronectin, collagen, or laminin (although peptides and antibodies have also been successfully conjugated) (*see* **Note 4**).
6. PBS, filtered and stored at room temperature.

2.4. Cell Culture and Media

1. Cells should be healthy and seeded onto the substrate under normal growth conditions at 5000 cells per 22-mm substrate (*see* **Note 5**).
2. If cells are to be transiently transfected, they should not be transfected on the hydrogels, but on conventional polystyrene dishes, and reseeded onto the hydrogels following recovery for an appropriate period of time (*see* **Note 6**).
3. An important control requires that a protein-free hydrogel be used for the top substrate. This substrate should not be exposed to serum prior to sandwiching.

2.5. Assembly of the Double-Hydrogel Substrate

1. Diamond-tip pen.
2. Stainless steel weight of approx 30 g.
3. A square piece of glass 2 cm × 2 cm × 5 mm, 4–5 g.
4. Sterile Pasteur pipets.

3. Methods

The double-hydrogel substrate involves sandwiching cells between two single-layer polyacrylamide hydrogels, which were described in detail for traction force detection *(9)*. The culture system provides cells with a flexible, porous support as in native tissues and with anchorage sites for integrins on both ventral and dorsal surfaces. We have previously shown that the top substrate makes initial contact with the thickest part of the cell, above the nucleus, which is sufficient to induce an elongated morphology in fibroblasts and a reduced number of actin stress fibers and focal adhesions (**Fig. 1B,C**), typical of fibroblasts in vivo and in collagen gels *(5–7)*. Individual extensions exert forces on either the bottom or the top substrate, but not both, because of their limited thickness *(8)*.

The double-hydrogel substrate offers several advantages over other 3D model systems as described in **Heading 1.**, including the ability to control the flexibility, porosity, and matrix protein coating. Because the cells do not penetrate into the gel, images may be collected from a limited number of focal planes without confocal laser scanning optics. The high optical quality offers excellent resolution for structures in live cells such as focal adhesions. However, these advantages also impose constraints to cell shape and migration, and the culture environment is not entirely isotropic as for cells embedded in collagen or Matrigel gels.

3.1. Activation of Cover Slips

Sheets of polyacrylamide hydrogels are cast between two cover slips. The bottom sheet is cast between a large cover slip (e.g., 45 × 50 mm, no. 1), which is to be mounted into a culture chamber, and a small 22-mm circular cover slip.

The top sheet may be prepared similarly, and then trimmed out of the large cover slip. The large cover slips need to be chemically activated to allow polyacrylamide to attach covalently; otherwise the gel will slide off during handling.

1. Clean the surface of the large cover slips and mark the side to be activated with a diamond pen.
2. Hold the cover slip to be activated with a pair of forceps and pass across the inner flame of a Bunson burner with the marked surface facing down to make the surface hydrophilic.
3. Smear a small volume of 0.1 N NaOH over the flamed surface of the cover slips using a Pasteur pipet. Allow the surface to air-dry.
4. Working in a fume hood, use a Pasteur pipet to smear the marked surface of the cover slips with 3-aminopropyltrimethoxysilane and incubate for 5 min.
5. Rinse the cover slips extensively with distilled water until the glass is clear.
6. Working in a fume hood, prepare a 0.5% solution of gluteraldehyde in PBS and pipet sufficient solution to cover the surface of the cover slips. Incubate for 30 min.
7. Wash the cover slips extensively in distilled water until the activated surface is clear; allow to air-dry.
8. Activated cover slips can be stored in a desiccator at room temperature for many weeks.

3.2. Preparation of Polyacrylamide Hydrogel Sheets

1. Activated cover slips are mounted into chamber dishes with the activated side facing up (*see* **Note 7**).
2. Although a much smaller volume would suffice, for convenience a 5-mL polyacrylamide solution is prepared at a final concentration of 5% acrylamide, 0.1% *bis*-acrylamide, and 10 mM HEPES. Stock acrylamide and *bis*-acrylamide solutions are mixed with distilled water, 1 M HEPES pH 8.5, and 50-µL latex beads to achieve the desired concentrations (*see* **Note 8**). Beads of different colors are used for top and bottom substrates (*see* **Note 2**).
3. The mixture is then degassed for 20 min to ensure reliable polymerization.
4. During degassing, collect reagents, pipets, and cover slips for the subsequent steps in order to cast the gel immediately following induction of polymerization.
5. After degassing add 30 µL of freshly prepared 10% w/v ammonium persulfate, and gently mix by swirling the container.
6. Complete the activation by adding 20 µL of TEMED and gently mix.
7. Working quickly, place a 15-µL drop of activated acrylamide solution onto the center of the activated cover slip and gently place a 22-mm round cover slip on the drop.
8. Quickly invert the sample to allow the fluorescent microbeads to settle to the surface of the gel away from the large cover slip.
9. Allow the gel to polymerize for at least 30 min.
10. After polymerization is complete, turn the sample right-side up and flood the gel with 50 mM HEPES, pH 8.5, to reduce surface tension between the small cover slip and the gel.

11. Remove the small cover slip with a pair of fine forceps immediately before conjugating matrix proteins to the gel (*see* **Note 9**). The small cover slip may be left in place if the gel is to be stored at 4°C.

3.3. Conjugation of Matrix Proteins to the Hydrogel

1. Prepare a fresh solution of 1 mM sulfo-SANPAH by first dissolving the solid into DMSO at 5% (w/v) concentration. Slowly add 50 mM HEPES, pH 8.5, while vortexing vigorously (*see* **Note 10**).
2. Cover the surface of the gel with 300 µL of solution and expose to UV light at a distance of 2.5 in. below the lamp tube for 6 min (*see* **Note 11**).
3. Rinse the solution from the gel surface with one wash of 50 mM HEPES, pH 8.5, and repeat **step 2**.
4. Quickly rinse two to three times with 50 mM HEPES, pH 8.5, and immediately add the solution of the matrix protein. Allow the conjugation reaction to proceed at 4°C overnight.
5. Rinse gently with PBS to remove excess matrix proteins and store in PBS at 4°C until ready for culturing.

3.4. Preparing Substrates for Culture

1. Polyacrylamide hydrogels must first be "sterilized" and equilibrated with media before culturing. Quasi-sterilization is performed by placing the gels covered with minimal PBS in a culture hood under the UV germicidal light for 10–20 min.
2. Remove PBS, add culture medium, and place in a CO_2 incubator for 30–60 min to allow the gel to swell to a steady state (*see* **Note 12**).
3. Freshly trypsinized cells are seeded onto the substrates at approx 5000 cells per 22 mm substrate.
4. Cells are typically given several hours to overnight to attach and spread.

3.5. Assembly of the Double-Hydrogel Substrate

1. Top substrates may be carved out of a cover slip already mounted in a culture chamber or from an unmounted large cover slip placed over a supporting frame (*see* **Note 13**). Using a diamond-tip pen, press down firmly and trace along the perimeter of the 22-mm circular hydrogel. Although the entire circle does not always come out intact, fragments of the circle may work equally well.
2. Place the top substrate into a Petri dish containing medium and return to the incubator until needed.
3. In preparation for the sandwich assembly, collect the 30-g weight, square weighing glass, forceps, media, and Pasteur pipets in the culture hood. The bottom substrate containing attached cells and the top substrate are then brought to the hood.
4. Working quickly, use a Pasteur pipet under vacuum to aspirate as much media as possible from the bottom substrate and to remove any large dangling pieces of gel that might affect the contact between the top and bottom substrates.
5. Use a pair of forceps to pick up the top substrate from the Petri dish. Aspirate off excess media and any dangling pieces of gel with a Pasteur pipet.

6. Gently place the cover slip with the top substrate over the cells. Make sure that the gel is facing down. Carefully place the square weighing glass over the sandwich followed by the 30-g stainless steel weight for 30 s.
7. Add a small amount of medium such that the surface is just below the top of the square glass. Remove the metal weight with a pair of forceps. The square weighing glass stays on the sandwich.
8. Return the culture to the incubator or begin the experiment (*see* **Note 14**).

3.6. Observing the Live Culture Under the Microscope

Because the distance between the top and bottom substrates is affected by the presence of debris, it is important that this distance be verified in the region of observation.

1. Cells are scanned at a low magnification (e.g., using a ×10 phase objective lens). 3T3 fibroblasts take a spindle or stellate shape within 120 min of application of the top substrate.
2. Switch to a dry objective lens of a higher magnification such as ×40. The distance between the top and bottom substrate is judged by focusing first on fluorescent beads on the top surface of the bottom gel, using fluorescence optics. The filter cube of the microscope is switched to match the color of beads in the top substrate. The focal plane is then brought up slowly until beads on the bottom surface of the top gel just come into focus. The distance is then estimated from the angular travel of the focusing knob, using a conversion factor provided by the microscope manufacturer.
3. 3T3 cells show an elongated morphology and maintain their motility when the substrate is separated by 3–6 µm. Cells in regions of a greater substrate distance show no response in morphology to the top gel, whereas those in regions of a smaller distance show impeded motility.

4. Notes

1. A variety of culture chambers may be used. One simple design, as described previously *(9)*, is made of Plexiglas with a 35-mm-diameter hole drilled into the center. The No.1 cover slip measuring 45 × 50 mm (Fisher Scientific) is mounted using silicone high vacuum grease. An alternative design uses a disposable 60-mm polystyrene dish with a 35-mm hole drilled through the middle using a cup saw. A 45-mm round cover slip (Labor Optik GmBH, Friedrichshofen, Germany) is then adhered using an optical adhesive (Norland Optical Adhesive 71, Norland Products Inc., Cranbury, NJ).
2. Red and green microspheres give optimal results for distinguishing top and bottom substrates. However, when cells are labeled with a fluorescent probe, such as green fluorescent protein, both substrates may be labeled with red beads and distinguished by the level of focusing.
3. We have found that sulfo-SANPAH activity may vary between lots. An alternative method is to use a classical carbodiimide such as EDC (1-ethyl-3-[3-dimethyl

aminopropyl]carbodiimide-HCl) on a composite gel of polyacrylamide/polyacrylic acid. This method has been described previously *(9)*.

4. Proteins to be conjugated by sulfo-SANPAH should not be dissolved in Tris buffer because it competes with the protein for binding to sulfo-SANPAH. Tris buffer may be replaced easily by dialysis. Some proteins, such as laminin, maintain better functions if bound noncovalently and indirectly to the substrate via a layer of polylysine conjugated with sulfo-SANPAH to the polyacrylamide.

5. A lower cell density allows for easier analysis of the cells and eliminates complications that arise from cell–cell contacts. However, the morphological change does occur even at higher densities.

6. We have experienced very low efficiency of transfections and high rate of cell death when using lipid-based transfection methods on cells adhered to the hydrogels. Transfection using the Amaxa Nucleofector before seeding cells onto the hydrogel gives optimal results (Amaxa, Gaithersburg, MD).

7. It is easier to work with the hydrogels when the cover slips are mounted to chamber dishes at an early stage of preparation. However, it is also possible to cast the gels on the unmounted cover slips and prepare the chamber later.

8. For a typical preparation of a 5%/0.1% gel, we use the following recipe: 625 µL 40% acrylamide, 250 µL 2% *bis*-acrylamide, 50 µL 1 *M* HEPES, 50 µL Fluouspheres, 4025 µL distilled water.

9. Leave the small 22-mm cover slip on top of the polyacrylamide if the gel is to be stored without conjugation, to better preserve the surface of the gel. The gel may be left in 50 m*M* HEPES, pH 8.5, at 4°C. The small cover slip may be removed using two pairs of fine-tipped forceps: one for bracing an edge and the second for lifting the opposite edge.

10. Sulfo-SANPAH is unstable in solution and should be used immediately after preparation. The reagent has a marginal solubility and may be better handled by first dissolving the solid in a small volume of DMSO. HEPES solution at room temperature may then be added while vortexing vigorously.

11. Photoactivation should cause the sulfo-SANPAH solution to darken considerably, but the extent of darkening can be inconsistent between experiments and between different lots of the reagent.

12. The physical properties of hydrogels vary with temperature and osmolarity. Therefore, it is very important to acclimate the gels to the actual culture condition before use.

13. Instead of cutting the top substrate out of a large cover slip, it may be prepared with two pieces of round 22-mm cover slips. However, it is often difficult to maintain the flatness of the gel along the edge when cast with small cover slips. This causes problems in controlling the space within the sandwich.

14. Depending on the experimental design, cell elongation may be observed from rounded and freshly attached cells or from cells already spread on the bottom substrate. The cells reach a similar steady state as soon as 2 h after applying the top substrate, although we typically culture the cells in the sandwich overnight before observation.

Acknowledgments

This work was supported by NIH GM-32476 awarded to Y.L.W.

References

1. Meier, S. and Hay, E. D. (1975) Stimulation of corneal differentiation by interaction between cell surface and extracellular matrix. I. Morphometric analysis of transfilter "induction." *J. Cell Biol.* **66,** 275–291.
2. Mauchamp, J., Chambard, M., Verrier, B., et al. (1987) Epithelial cell polarization in culture: orientation of cell polarity and expression of specific functions, studied with cultured thyroid cells. *J. Cell Sci.* **8,** 345–358.
3. Cook, J. R., Crute, B. E., Patrone, L. M., Gabriels, J., Lane, M. E., and Van Buskirk, R. G. (1989) Microporosity of the substratum regulates differentiation of MDCK cells in vitro. *In Vitro Cell Dev. Biol.* **25,** 914–922.
4. Schmeichel, K. L. and Bissell, M. J. (2003) Modeling tissue-specific signaling and organ function in three dimensions. *J. Cell Sci.* **116,** 2377–2388.
5. Grinnell, F. (2003) Fibroblast biology in three-dimensional collagen matrices. *Trends Cell Biol.* **13,** 264–269.
6. Cukierman, E., Pankov, R., Stevens, D. R., and Yamada, K. M. (2001) Taking cell-matrix adhesions to the third dimension. *Science* **294,** 1708–1712.
7. Friedl, P. and Brocker, E. B. (2000) The biology of cell locomotion within three-dimensional extracellular matrix. *Cell. Mol. Life Sci.* **57,** 41–64.
8. Beningo, K. A., Dembo, M., and Wang, Y.-L. (2004) Responses of fibroblasts to anchorage of dorsal extracellular matrix receptors. *Proc. Natl. Acad. Sci. USA* **101,** 18,024–18,029.
9. Beningo, K. A., Lo, C.-M., and Wang, Y.-L. (2002) Flexible polyacrylamide substrata for the analysis of mechanical interactions at cell-substratum adhesions. *Methods Cell Biol.* **69,** 325–339.

15

In Vitro Actin Assembly Assays and Purification From *Acanthamoeba*

J. Bradley Zuchero

Summary

The actin cytoskeleton is essential to all eukaryotic cells. In addition to playing important structural roles, assembly of actin into filaments powers diverse cellular processes, including cell motility and endocytosis. Actin polymerization is tightly regulated by various cofactors, which control spatial and temporal assembly of actin as well as the physical properties of these filaments. Development of an in vitro model of actin polymerization from purified components has allowed for great advances in determining the effects of these proteins on the actin cytoskeleton. The pyrene actin assembly assay is a powerful tool for determining the effect of a protein on the kinetics of actin assembly, either directly or as mediated by proteins such as nucleators or capping factors. In addition, fluorescently labeled phalloidin can be used to visualize the filaments that are created in vitro to give insight into how these proteins influence actin filament superstructure.

Key Words: Actin polymerization assay; *Acanthamoeba*; fluorimetry; pyrene; cytoskeleton; phalloidin.

1. Introduction

Over the past few decades dozens of actin-binding proteins have been discovered, and there is a rapidly growing body of work focused on dissecting how each of these proteins regulates the kinetics and morphology of actin structures *(1)*. Actin dynamics can be divided into three steps, each of which is specifically regulated. Polymerization is a nucleation–condensation reaction *(2)*, which means that the first step (nucleation) is slow, and the second step (elongation) proceeds quickly once a stable nucleus is formed *(3)*. Nucleation refers to the assembly of actin monomers into a stable trimer, which is usually an extremely unfavorable reaction. As the filament elongates, incorporated actin monomers

From: *Methods in Molecular Biology, vol. 370:*
Adhesion Protein Protocols, Second Edition
Edited by: A. S. Coutts © Humana Press Inc., Totowa, NJ

hydrolyze their ATP to ADP, making filaments more susceptible to depolymerization, the third step in the cycle. All steps of this cycle are regulated by actin-binding proteins *(1)*. The rate of nucleation is controlled by three distinct classes of actin nucleators: the Arp2/3 complex *(4)*, formins, and spire *(5,6)*. Growth, maintenance, and structural integrity of filaments are controlled by classes of proteins that cap, bundle, sever, or disassemble filaments.

The primary in vitro assay used to study the effects of actin-binding proteins on actin polymerization is the pyrene actin assembly assay, wherein the environmentally sensitive fluorophore pyrene indicates the polymerization state of actin *(7)*. Pyrene-labeled actin incorporated in a filament fluoresces significantly greater than a labeled monomer in solution *(8)*. Increase in fluorescence over time is directly proportional to the increase in filamentous actin *(9)*.

Actin polymerization assays are conducted using buffer conditions similar to those in vivo. Importantly, actin requires magnesium-ATP in its nucleotide pocket and physiological concentrations of potassium for rapid polymerization at micromolar concentrations. Actin is stored in Buffer A—which contains calcium instead of magnesium and lacks potassium—to prevent polymerization. Immediately before the assay, calcium is exchanged for magnesium by adding ethylene glycol tetraacetic acid (EGTA) and magnesium. Polymerization is triggered by the addition of buffer containing potassium *(10)*.

There are two standard sources of actin for biochemical studies: *Acanthamoeba castellani* and rabbit skeletal muscle. Actin from ameba is cytoplasmic and is a good model for cytoplasmic actins from other organisms *(11)*. Skeletal actin does not bind to some actin-binding proteins *(12)*. Furthermore, *Acanthamoeba* is an excellent source of actin-binding proteins, such as the Arp2/3 complex *(13,14)*. Although it is best to use actin and actin-binding proteins from the same organism and tissue, actin is highly conserved, and in most cases ameba actin can be used in conjunction with proteins from other organisms *(6,10)*.

Although the pyrene assay is very powerful for measuring kinetics of actin assembly, it does not reveal the structure of the actin filament networks it produces. There are two commonly used techniques that address this: (1) fixing and staining polymerization reactions with fluorescently labeled phalloidin *(6)* and (2) electron microscopy *(15)*. Fixing actin with labeled phalloidin, detailed below, is a straightforward method that can be done in parallel with pyrene actin experiments.

2. Materials

2.1. Culture of A. castellani

1. Ameba medium: 7.5 g/L proteose peptone, 7.5 g/L yeast extract, 15 g/L glucose, 0.2 mM methionine, 3 mM KH$_3$PO$_4$, 10 μM CaCl$_2$, 1 μM FeCl$_3$, 0.1 mM MgSO$_4$, 1 mg/L thiamine, 0.2 mg/L biotin, 0.01 mg/L vitamin B$_{12}$. Autoclave before use.

2. Wash buffer: 10 m*M* Tris-HCl, pH 8.0, 150 m*M* NaCl.

2.2. *Actin Purification From* A. castellani

1. Extraction buffer: 10 m*M* Tris-HCl, pH 8.0, 11.6% (w/v) sucrose, 1 m*M* EGTA, 1 m*M* ATP, 5 m*M* dithiothreitol (DTT) (Roche Applied Science, Mannheim, Germany), 30 mg/L benzamidine (Sigma-Aldrich, St. Louis, MO), 5 mg/L pepstatin A (Sigma-Aldrich), 10 mg/L leupeptin (Sigma-Aldrich), 40 mg/L soybean trypsin inhibitor (Sigma-Aldrich), 1 m*M* phenylmethylsulfonylfluoride (PMSF; Sigma-Aldrich). The first three ingredients may be combined a day in advance and stored at 4°C. Add the final ingredients fresh, pH to 8.0, and bring up to volume with cold water (*see* **Note 1**). Add PMSF immediately before use.
2. Pepstatin A (1000X): 5 mg/mL pepstatin A (Sigma-Aldrich) in dimethylsulfoxide; store at −20°C.
3. Protease inhibitor cocktail (100X): 1 mg/mL leupeptin and 4 mg/mL soybean trypsin inhibitor in water; store at −20°C.
4. PMSF (Sigma-Aldrich): 200 m*M* in ethanol; store at −20°C. Before use, warm to 25°C with rocking to resuspend. The half-life of PMSF in aqueous solution is approx 30 min, and so it is always added to buffers immediately before use.
5. 100 m*M* ATP, pH 7.0, dissolved in water.
6. Column buffer: 10 m*M* Tris-HCl, pH 8.0, 0.2 m*M* CaCl$_2$, 30 mg/L benzamidine, 0.5 m*M* ATP, 0.5 m*M* DTT. Add last three ingredients fresh and bring buffer up to volume with cold water. Adjust buffer to pH 8.0 while at 4°C, just before use.
7. Diethylaminoethyl cellulose (DEAE) (DE52, Whatman, Maidstone, England). After use, cycle with 0.5 *M* NaOH and HCl to wash, and store in Tris base, pH 8.0, 0.2% (v/v) benzalkonium chloride at 4°C indefinitely (*see* **Note 2** for details on washing DEAE).
8. 1 *M* Tris base (Trizma, Sigma-Aldrich).
9. High-salt buffer: 600 m*M* KCl, 10 m*M* Tris-HCl, pH 8.0, 0.2 m*M* CaCl$_2$, 30 mg/L benzamidine, 0.5 m*M* ATP, 0.5 m*M* DTT. Prepared as extraction buffer (**step 1**).
10. 200 m*M* MgCl$_2$.
11. Tris(2-carboxyethyl)phosphine hydrochloride (TCEP; Invitrogen, Carlsbad, CA): prepare 0.5 *M* stock in cold water and pH to 7.0 with KOH.
12. Buffer A (1X): 2 m*M* Tris-HCl, pH 8.0, 0.2 m*M* ATP, 0.5 m*M* TCEP, 0.1 m*M* CaCl$_2$. Bring pH to 8.0 with KOH, and bring up to volume with cold water (*see* **Note 3**).
13. KMEI buffer (10X): 500 m*M* KCl, 10 m*M* MgCl$_2$, 10 m*M* EGTA, 100 m*M* imidazole HCl, pH 7.0 (*see* **Note 4**).
14. Gel filtration column: flex column (25 mm × 100 cm, Fisher Scientific, Pittsburgh, PA).
15. Sephacryl S-200 HR resin (GE Healthcare, Little Chalfont, UK). Resin comes stored in 20% (v/v) ethanol and will swell when fully hydrated. To avoid cracking the resin, equilibrate in buffer before decanting into gel filtration column. Pack the column by flowing one to two column volumes of buffer at a faster rate than will be used in gel filtration.

2.3. Pyrene Labeling of Actin

1. Actin: If actin is from *Acanthamoeba*, it must be gel filtered before labeling. Skeletal actin does not have this requirement.
2. KMEI buffer (10X): *see* **Subheading 2.2., item 13**.
3. Dialysis buffer: 50 m*M* KCl, 1 m*M* MgCl$_2$, 1 m*M* EGTA, 10 m*M* imidazole HCl, pH 7.0, 0.2 m*M* ATP.
4. Pyrenyl iodoacetamide (Invitrogen): 10 m*M* stock solution in DMF, stored at −20°C in the dark.
5. Buffer A (1X): *see* **Subheading 2.2, step 12**.
6. Dithiothreitol (DTT): 1 *M* stock in water.

2.4. Pyrene Actin Assembly Assay

1. Buffer A (1X): *see* **Subheading 2.2, step 12**.
2. KMEI buffer (10X): *see* **Subheading 2.2, step 13**.
3. ME buffer (10X): 0.5 m*M* MgCl$_2$, 2 m*M* EGTA.
4. Ethanol, high-performance liquid chromatography grade, for washing cuvets.
5. Vakuwash cuvet washer (Fisher Scientific).

2.5. Microscopy of Labeled Actin Filaments

1. Reagents for pyrene actin assembly assay (**Subheading 2.4.**).
2. Alexa-fluor 488 Phalloidin (Invitrogen): make stocks in methanol, to be in molar excess to actin in reaction. For example, 5 µL of a 6.6 µ*M* stock of phalloidin was used to stabilize an equal volume of 4 µ*M* actin *(6)*. Store stocks at −20°C for several months. TRITC phalloidin (Sigma) also works well.
3. Wide-bore pipet tips: cut pipet tips with a sterile razor, and be consistent.
4. Poly-L-lysine: prepare a 1 mg/mL stock in water.
5. Poly-L-lysine-coated cover slips: clean slides by sonication first in 1 *M* KOH, then in ethanol. Rinse in water and air-dry. Spot 1 mg/mL poly-L-lysine on parafilm and float cover slips on drops (500 µL) for a few minutes; rinse with water, and use within a few days.
6. Clear nail polish.

3. Methods

3.1. Culture of A. castellani

1. Passage ameba in 10-mL culture tubes with moderate bubbling of humidified air at 25°C *(16,17)*.
2. Fill beveled Fernbach flasks (3 L) with 1 L ameba medium and inoculate with cultures. Grow with moderate shaking at 25°C until growth plateaus (*see* **Note 5**). Use Fernbachs to inoculate 14-L carboys outfitted with bubblers. Practice sterile technique to avoid contaminating the cultures, which grow without any antibiotic.
3. When cultures plateau, harvest by centrifugation at 4500*g* (4000 rpm in a Sorvall RC-3B) for 5 min. Refill tubes and centrifuge again until all ameba have been pelleted.

4. Wash ameba with wash buffer. Resuspend cell pellets in the smallest volume of buffer possible. Pool these and centrifuge as above. In this way all ameba are transferred into one or two tubes. During the pairing-down of tubes, weigh ameba to get approx 500 g (enough for actin purification) into a single tube.

3.2. Purification of Actin From A. castellani

1. (Example volumes given are for purification from 500 g ameba. Buffer and resin volumes can be scaled linearly as needed.) Resuspend pelleted ameba in 1 L cold extraction buffer. If ameba are frozen, thaw in cold extraction buffer (*see* **Note 6**). After resuspension, conduct all subsequent steps at 4°C (*see* **Note 7**).
2. Lyse cells by nitrogen decompression in a Parr Bomb (Parr Instrument Company, Moline, IL). Rinse the Parr Bomb with water and cool on ice. Transfer cells into the Parr Bomb and pressurize with nitrogen gas at 400 psi for 5 min. To lyse, release cells from the Parr Bomb slowly, while maintaining constant pressure.
3. To remove cellular debris, centrifuge lysate at more than $5000g$ (7000 rpm in a GS-3 rotor) for 10 min at 4°C. Centrifuge supernatant from this spin at more than $100,000g$ (38,000 rpm in a Beckman Ti45 rotor) for 2 h at 4°C.
4. Equilibrate 500 g DEAE resin in 2 L of column buffer. To equilibrate large volumes of resin, mix resin and buffer in beaker, then pour over a Büchner funnel fitted with a Whatman filter 541 (ashless) and pull vacuum until buffer has passed through but the resin is still damp. It is essential that the pH of the resin is 8.0 before use.
5. After the second spin, siphon off the lipid layer at the very top of each tube. Decant supernatants into a rinsed glass beaker, on ice. Mix supernatant with approx 450 mL equilibrated DEAE resin, and stir for 30–60 min at 4°C to batch bind.
6. In the cold room, pour remaining approx 50 mL equilibrated DEAE resin into a glass column (diameter 5 cm; height 60 cm) and let resin settle and excess liquid drip through. Gently pour DEAE slurry above this, and collect flow-through (*see* **Note 8**). Wash resin with 1 L column buffer, pumped at 6 mL per min.
7. To elute actin, use a gradient maker to run a linear gradient from 0 to 600 mM KCl using 1 L column buffer (low-salt buffer) and 1 L high-salt buffer, pumped at 3 mL per min. Actin normally elutes as a broad (~200 mL) peak between 300 and 400 mM KCl. Keep actin fractions in the cold room overnight, covered. (*See* **Note 2** regarding recycling of DEAE resin.)
8. The next day, move fractions to 25°C to speed up polymerization. Actin will polymerize in a few hours. Detect polymerized actin fractions by tipping fraction tubes sideways; if there is sufficient polymerized actin, the fraction will be a gel.
9. Pool actin fractions into a beaker. Add $MgCl_2$ to 2 mM and ATP to 1 mM. Stir actin gently for 30 min at 25°C to complete polymerization.
10. Centrifuge actin at $1000g$ (3000 rpm in a Sorvall GSA rotor) for 5 min at 4°C. Pour off supernatant, and wash with 1X KMEI buffer. Centrifuge at more than $100,000g$ (38,000 rpm in a Ti45 rotor) for 2 h at 4°C. Discard supernatant and rinse the pellet (polymerized actin) quickly with Buffer A. Scrape actin into a Dounce homogenizer.

Homogenize 5–10 times with a "loose" pestle in Buffer A to 40–80 mL final volume without creating air bubbles.

11. Dialyze into Buffer A for 2–3 d at 4°C to depolymerize actin. Change dialysis buffer daily. Actin can be stored in dialysis at 4°C for several weeks.
12. Equilibrate an S-200 column with Buffer A. Centrifuge 15–20 mL depolymerized actin at more than 100,000g (38,000 rpm in a Ti45 rotor) for 2 h at 4°C. Load actin onto S-200 column and gel filter in Buffer A, collecting approx 3-mL fractions.
13. Determine the concentration of unlabeled actin by measuring absorbance at 290 nm, using an early fraction from the column to blank the spectrophotometer (*see* **Note 9**). The concentration of actin is 38.5 µM per O.D. 290, assuming a 10-mm pathlength.
14. Gel-filtered actin can now be pyrene labeled or used unlabeled in polymerization assays (**Fig. 1A**). For pyrene labeling, use the entire second (major) peak. Dark actin for fluorimetry is taken from the back end (later fractions) of this peak. The front end of the peak contains actin dimers, which will interfere with kinetic measurements *(10)*.
15. Assay the purity of actin by sodium dodecyl sulfate–polyacrylamide gel electrophoresis and Coomassie staining. Actin is 42 kDA and should appear as a single band with little to no breakdown.

3.3. Pyrene Labeling of Actin

1. Pool 30–80 mg of actin and measure concentration. Dilute actin to 1.1 mg/mL (26.4 µM; because of the limited solubility of pyrene iodoacetamide, the final concentration of actin should be 1 mg/mL) in Buffer A.
2. Polymerize actin by adding one-tenth the final volume of 10X KMEI buffer and let stand at 25°C. Actin will be completely polymerized within a few minutes; look for bubbles in the actin solution that do not move.
3. To remove reducing reagent for labeling, carefully transfer polymerized actin into dialysis tubing, and dialyze into dialysis buffer for a few hours at 4°C.
4. Transfer actin to a beaker and measure the volume. Add 4–7 mol pyrenyl iodoacetamide per mol actin while stirring. Cover the beaker to keep dark, and gently stir overnight at 4°C. Quench reaction by adding DTT to 10 mM.
5. Centrifuge labeled actin at 2000g (5000 rpm in a Beckman JA-20) rotor for 5 min at 4°C to pellet precipitated dye.
6. Centrifuge supernatant at more than 100,000g (38,000 rpm in a Ti45 rotor) for 2 h at 4°C to pellet actin filaments.
7. Homogenize the pellet with a Dounce homogenizer, as above, in Buffer A, into a final concentration of about 2–6 mg/mL (48–144 µM) (*see* **Note 10**).
8. Dialyze into Buffer A at 4°C using at least two changes of buffer over at least 2 d to depolymerize actin.
9. Centrifuge depolymerized pyrene actin at more than 100,000g (38,000 rpm in a Ti45 rotor) for 2 h at 4°C to remove remaining actin filaments.
10. Gel filter supernatant over S-200 resin equilibrated with Buffer A, as before.

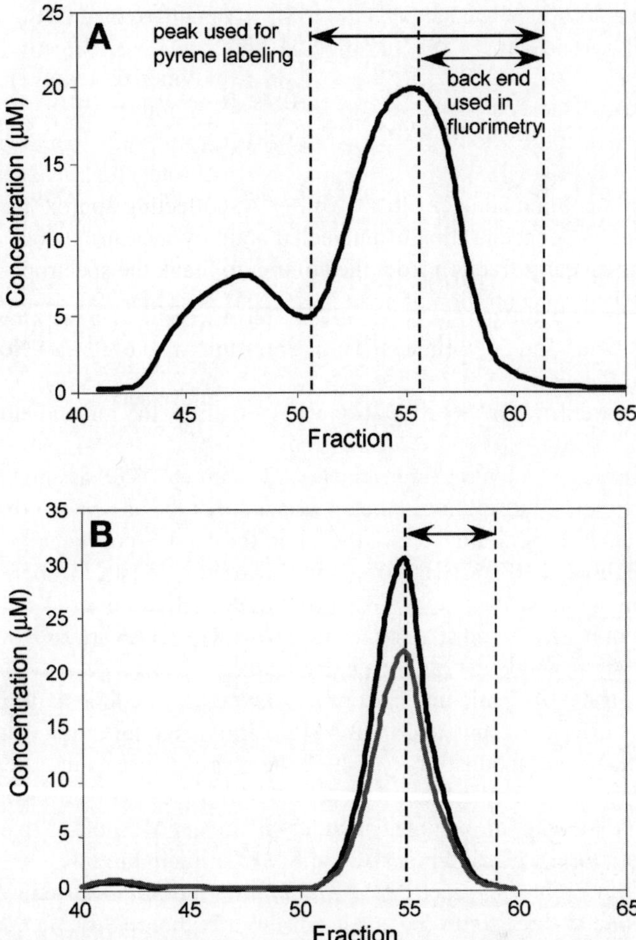

Fig. 1. Gel filtration of actin from *Acanthamoeba castellani.* **(A)** Depolymerized actin was gel filtered over S-200 resin in Buffer A. Absorbance at 290 nm was measured and concentration of actin plotted for each fraction. The first peak contains contaminants, including α-actinin. The second (major) peak is actin. It is important to use only the back end of the peak for fluorimetry *(20).* **(B)** Pyrene-labeled actin was gel purified over S-200 resin in Buffer A. Absorbances at 290 and 344 nm were measured, and concentrations of total actin (black line) and pyrene actin (gray line) were plotted. Percent labeled equals the ratio of pyrene actin to total actin. As above, use the back end of the peak for fluorimetry.

11. Determine the concentration and percent labeling by measuring absorbance at 290 and 344 nm. To correct for pyrene absorption at 290 nm:

$$A_{290(\text{corrected})} = A_{290(\text{measured})} - 0.127 \times A_{344} \qquad (1)$$

The concentration of total actin is thus 38.5 μM per corrected OD_{290}. The concentration of pyrene actin is 45.0 μM per OD_{344}. To calculate percent labeling:

$$\text{Percent labeled} = \frac{[\text{pyrene actin}]}{[\text{total actin}]} \tag{2}$$

12. Store pyrene actin fractions on ice, in the dark, in the cold room. Pyrene actin is more stable than unlabeled actin and can be used for several months.

3.4. Pyrene Actin Assembly Assay

1. Mix unlabeled actin, pyrene actin, and Buffer A to achieve a working stock that is 5–10% labeled and 5–20 times the concentration to be used (*see* **Note 11**). Store on ice, in the dark.
2. Set up the fluorimeter. Set excitation wavelength to 365 nm and emission wavelength to 407 nm.
3. Assemble reagents by the fluorimeter. (All volumes listed assume 100-µL reaction.) Prealiquot buffers into labeled Eppendorf tubes to ensure that concentrations of each reagent are exactly the same for each experimental condition: (1) 10-µL aliquots of 10X KMEI buffer and (2) aliquots of 10X ME buffer (one-tenth the volume of actin added to the reaction). Make a dilution series of the protein to be tested in Buffer A and store on ice (*see* **Note 12**). Clean cuvets extensively with water, ethanol, and water, and wipe dry with lens paper.
4. Set a stopwatch to 3 min and start time. Immediately, add actin to prealiquotted 10X ME buffer to exchange Ca^{2+} for Mg^{2+}. Exchange for 2 min exactly.
5. During this 2-min incubation, assemble the rest of the reagents: Add proteins to be tested to 10 µL prealiquotted 10X KMEI buffer (*see* **Note 13**). Add Buffer A to bring the total reaction volume (including actin and ME buffer) to 100 µL. Set a pipetteman to 110 µL, and pipet up the KMEI–protein mixture.
6. After 2-min exchange, pipet KMEI–protein mixture into the actin. Triturate the reaction one to three times (be consistent) to mix thoroughly, then pipet into the cuvet, being careful to avoid bubbles. Load the cuvet into the fluorimeter and activate data collection. Record dead time: the time from addition of the KMEI–protein mixture to actin until the second the first data point is collected. Always collect complete data sets until pyrene fluorescence plateaus. Try the same condition at least three times to ensure reproducibility. (*See* **Note 14** for tips on improving data.)
7. Every time the fluorimeter is turned on, or a different concentration of actin is used, optimize settings to maximize signal while eliminating photobleaching (*see* **Note 15**). Polymerize actin as described, and monitor the plateau over the course of 1 h, maximizing the signal that causes no appreciable drop in plateau fluorescence over time. This is usually achieved by adjusting the diaphragm between the light source and the cuvet (*see* **Note 16**).
8. **Figure 2** illustrates polymerization data and how they are analyzed. Plot fluorescence (arbitrary units) as a function of time (in seconds). Add back dead time to *x*-axis data (time). Subtract baseline fluorescence from all *y*-axis data. Baseline fluo-

rescence is fluorescence at time zero, assuming the reaction is slow enough to catch. If the conditions used do not alter the critical concentration of actin (i.e., protein being tested does not cap actin filaments or sequester monomers), normalize data by dividing all y-axis data by the fluorescence at plateau. *See* **ref.** *10* for methods to determine critical concentration change.

9. Reaction half-times are convenient metrics to measure (**Fig. 2C,** inset). Replot the linear region of the curve surrounding half-maximal fluorescence, and fit a linear equation to these points. Use this equation to solve for half time (the value of x where $y = 0.5$, if data are normalized).

3.5. Microscopy of Labeled Actin Filaments

1. Polymerize unlabeled actin using conditions described above in an Eppendorf tube. Allow reaction to come to plateau.
2. In the meantime, desiccate Alexa-Fluor 488 phalloidin in a speed vacuum to remove all methanol. Labeled phalloidin should be superstoichiometric to actin.
3. Pipet 5 µL of polymerized actin—using a wide-bore pipet tip—into the tube containing desiccated phalloidin to stabilize actin filaments.
4. Add 45 µL of 1X KMEI buffer to dilute the reaction. Dilute to low nanomolar concentrations to resolve individual filaments.
5. Pipet 10 µL of the dilutions onto slides and cover with poly-L-lysine cover slips. Seal slides with clear nail polish.
6. Once nail polish dries, image filaments with an epifluorescence microscope.

4. Notes

1. All solutions should be prepared in double-distilled water with resistivity of 18.2Ω. Cold water refers to double-distilled water at 4°C.
2. More complete protocol: dry DEAE resin (vacuum resin until just damp; not white). Transfer to beaker and add water to 4 L. Add solid NaOH to 0.5 *M* and stir 5 min at 25°C. Dry and wash with 4 L water. Repeat NaOH wash; this time stir for 30 min. Dry and wash with 8 L water. Resuspend in 2870 mL water and add 130 mL concentrated HCl. Stir 30 min at 25°C. Dry and wash with 8 L water. Resuspend in 3–4 L water and add unbuffered 1 *M* Tris base until pH is 8.0. Add 0.2% (v/v) benzalkonium chloride and store at 4°C.
3. Make Buffer A as a 100X stock without $CaCl_2$—calcium–ATP precipitates at this concentration—and freeze at −20°C. Add $CaCl_2$ during dilution to 1X, and aliquot in 15-mL conical tubes and freeze at −20°C. Once thawed, one aliquot may be used for up to 1 wk.
4. Imidazole is best stored in the dark, at 4°C, as a 1–2 *M* stock solution. Store 10X KMEI buffer for fluorimetry in the dark at 4°C. Discard when autofluorescence of the buffer becomes noticeable during fluorimetry.
5. Assay growth by absorbance at 600 nm. A typical OD 600 of ameba for purifying actin is 6. OD 600 values greater than 1.0 are out of the linear range in most cases, so dilute the ameba before measuring. Alternatively, ameba can be grown in a fermentor.

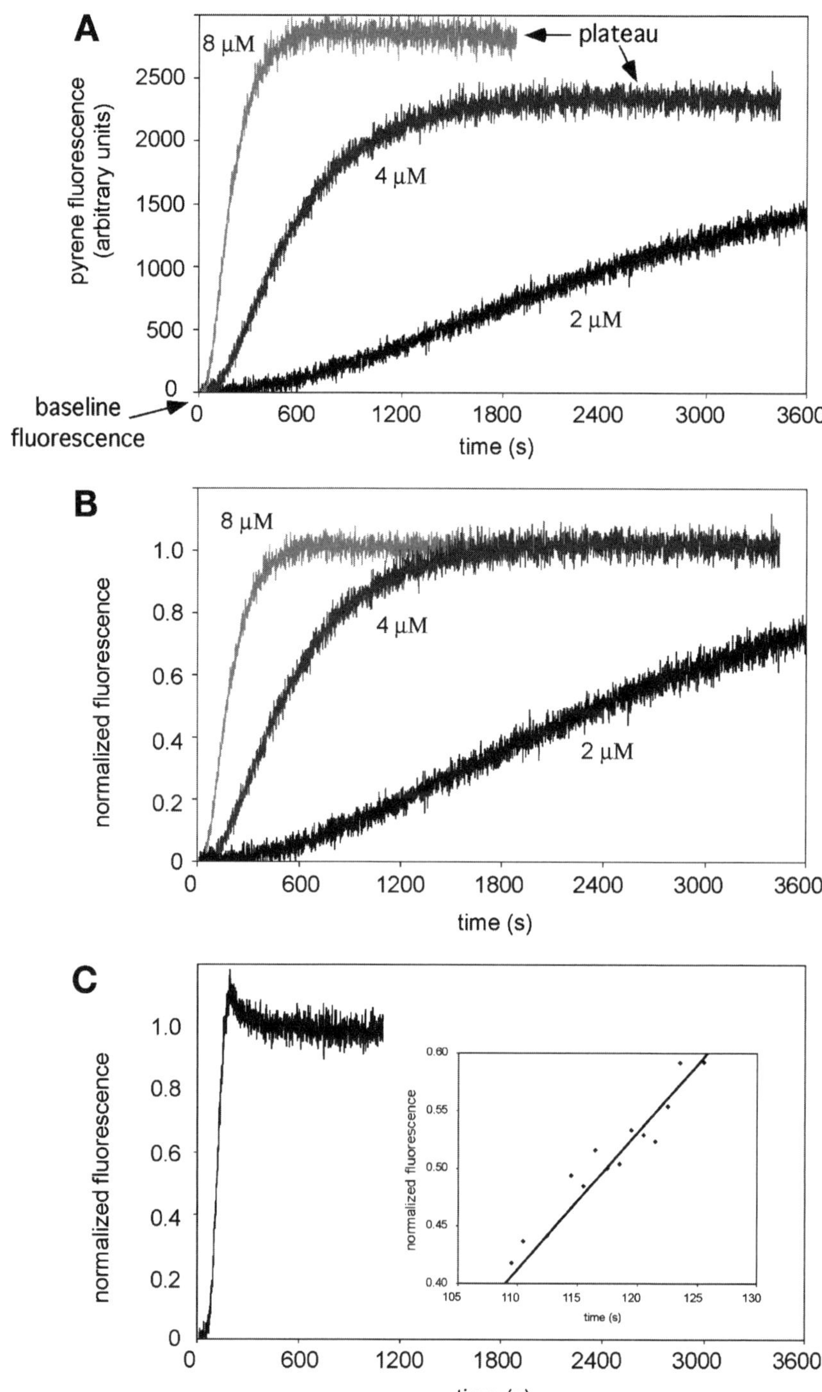
A
8 μM
plateau
2500
2000
4 μM
1500
1000
500
2 μM
0
baseline
fluorescence
0
600
1200
1800
2400
3000
3600
time (s)
pyrene fluorescence
(arbitrary units)

B
8 μM
1.0
0.8
4 μM
0.6
0.4
2 μM
0.2
0
0
600
1200
1800
2400
3000
3600
time (s)
normalized fluorescence

C
1.0
0.8
0.6
0.4
0.2
0
0
600
1200
1800
2400
3000
3600
time (s)
normalized fluorescence
0.60
0.55
0.50
0.45
0.40
105
110
115
120
125
130
time (s)
normalized fluorescence

6. We have noticed inconsistencies with actin purified from ameba that have been frozen for longer than 6 mo.
7. The first day of the actin purification protocol is significantly long and is aided by making buffers and gathering equipment and reagents the day before. Once started, the entire purification takes 5–7 d, with 2–3 d waiting for the actin to depolymerize in the middle.
8. To eliminate bubbles when decanting resin slurry into the column, pour it slowly down the outside of a Pasteur pipet or glass stir rod onto the inside wall of the column. Do not let the resin drip directly into the column.
9. Actin requires ATP, which absorbs significantly less at 290 nm than at 280 nm. To limit the time that actin fractions are out of the cold room, it is practical to reuse pipets and not wash the cuvet between each reading. To avoid contaminating monomeric actin with actin dimers, start at the highest fraction and take readings backwards.
10. The critical concentration of calcium-actin is 150 μM *(10)*. Actin will not depolymerize above this concentration.
11. Concentrations of both dark actin and pyrene actin stocks must be measured. Because some proteins do not bind as tightly to pyrene actin as to dark actin, the actin used in fluorimetry is normally only 5–10% pyrene actin *(10)*. To calculate how much pyrene actin and dark actin to combine to yield a given percent labeling and a given concentration, use the following equations:

Let L_i = starting percent labeling of pyrene actin (100% labeled = 100)
Let C_p = starting concentration of pyrene actin (μM)
Let C_d = starting concentration of dark actin (μM)
Let L_f = final percent labeled (i.e., 5–10)
Let C_f = final concentration of actin (μM)
Let V_f = final volume of actin (μL)

$$\text{volume pyrene actin } (\mu L) = \frac{L_f \cdot C_f \cdot V_f}{L_i \cdot C_p} \tag{3}$$

$$\text{volume dark actin } (\mu L) = \left(V_f - \frac{\text{volume pyrene action} \cdot C_p}{C_f}\right) \cdot \frac{C_f}{C_d} \tag{4}$$

Bring up to volume (V_f) with Buffer A.

Fig. 2. *(Opposite page)* Pyrene actin assembly assay. **(A)** Actin was polymerized as described; curves show spontaneous nucleation of 2 μM (black), 4 μM (dark gray), and 8 μM (light gray) actin. These data have been zeroed by subtracting baseline fluorescence from all fluorescence values. **(B)** Data from above were normalized by dividing all fluorescence values by average fluorescence at plateau. **(C)** 2 μM actin was polymerized in the presence of 10 nM Arp2/3 complex and 5 nM *Listeria* ActA. Inset: To calculate time to half-maximal fluorescence ($t_{1/2}$), the linear region of the curve in **(C)** spanning 0.5 was replotted and a linear trendline was fit to the data. Solving for time (x) where $y = 0.5$ yields $t_{1/2}$.

12. Because ATP absorbs significantly at 280 nm, proteins other than actin should not be in Buffer A for quantification by spectrophotometry.
13. It may be desirable to premix proteins before starting actin exchange to allow them to come to equilibrium. Although not recommended for detailed kinetic analysis, it is possible to add *Escherichia coli* lysate containing expressed protein to pyrene actin assembly assays as a first-pass functional test. Uninduced (control) lysate should have no effect on actin polymerization.
14. There are many factors to consider for collecting meaningful and reproducible fluorimetry data. To list key points:
    a. Always do an actin-alone (control) reaction to test the actin first.
    b. The assay is very sensitive to the concentration of potassium, so keep that constant in all reactions. In the event that a protein component requires potassium, adjust the molarity of KCl in 10X KMEI buffer to keep the final concentration of potassium in the reaction at 50 m*M*.
    c. Exchange in 10X ME buffer for exactly 2 min and be consistent.
    d. Eliminate photobleaching.
    e. Always collect complete data sets—until pyrene fluorescence plateaus.
    f. Wash cuvets extensively between each reaction, because polymerized actin remaining in the cuvet will nucleate new filaments in the next run.
    g. Do your best to avoid bubbles, which both scatter light and cause oxidative damage to proteins.
    h. Store proteins on ice, but ensure that all components are at 25°C at the start of the reaction. Preincubating tubes in a water bath works well.
    i. When expressing glutathione-*S*-transferase (GST)-tagged recombinant proteins, it is critical to remove the GST, because it will mediate the dimerization of the fusion protein. For example, Arp2/3 activators display faster kinetics when the GST remains uncleaved, which is thought to be an artifact of this dimerization *(18,19)*.
15. Photobleaching is a complex phenomenon that cannot be corrected for mathematically *(10)*. If the overall signal is too low after eliminating photobleaching, try widening emission slits, increasing the voltage of the photomultiplier, or increasing pyrene-doping of actin, and again measuring photobleaching after the change.
16. We polymerize the concentration of actin that will be used in the experiments in the presence of Arp2/3 and an activator (like *Listeria* ActA), which comes to plateau very quickly. It is also possible to polymerize a high concentration of actin, and then dilute down to working concentration and measure in the fluorimeter.
17. First, to determine when a given condition will plateau, perform pyrene actin assembly assay as above. For microscopy, reaction volumes can be scaled down to less than 100 μL.

Acknowledgments

The author would like to thank Professor R. Dyche Mullins, Dr. Margot Quinlan, Orkun Akin, and Christopher Campbell for technical assistance and critical reading of the manuscript. This work was supported by an award from the American Heart Association.

References

1. Pollard, T. D. and Borisy, G. G. (2003) Cellular motility driven by assembly and disassembly of actin filaments. *Cell* **112,** 453–465.
2. Oosawa, F. and Kasai, M. (1962) A theory of linear and helical aggregations of macromolecules. *J. Mol. Biol.* **4,** 10–21.
3. Frieden, C. (1983) Polymerization of actin: mechanism of the Mg^{2+}-induced process at pH 8 and 20 degrees C. *Proc. Natl. Acad. Sci. USA* **80,** 6513–6517.
4. Welch, M. D. and Mullins, R. D. (2002) Cellular control of actin nucleation. *Annu. Rev. Cell Dev. Biol.* **18,** 247–288.
5. Baum, B. and Kunda, P. (2005) Actin nucleation: spire—actin nucleator in a class of its own. *Curr. Biol.* **15,** R305–R308.
6. Quinlan, M. E., Heuser, J. E., Kerkhoff, E., and Mullins, R. D. (2005) Drosophila spire is an actin nucleation factor. *Nature* **433,** 382–388.
7. Machesky, L. M., Mullins, R. D., Higgs, H. N., et al. (1999) Scar, a WASp-related protein, activates nucleation of actin filaments by the Arp2/3 complex. *Proc. Natl. Acad. Sci. USA* **96,** 3739–3744.
8. Cooper, J. A., Walker, S. B., and Pollard, T. D. (1983) Pyrene actin: documentation of the validity of a sensitive assay for actin polymerization. *J. Muscle Res. Cell Motil.* **4,** 253–262.
9. Pollard, T. D. (1983) Measurement of rate constants for actin filament elongation in solution. *Anal. Biochem.* **134,** 406–412
10. Mullins, R. D. and Machesky, L. M. (2000) Actin assembly mediated by Arp2/3 complex and WASP family proteins. *Methods Enzymol.* **325,** 214–237.
11. Gordon, D. J., Eisenberg, E., and Korn, E. D. (1976) Characterization of cytoplasmic actin isolated from *Acanthamoeba castellanii* by a new method. *J. Biol. Chem.* **251,** 4778–4786.
12. Blanchoin, L. and Pollard, T. D. (1999) Mechanism of interaction of *Acanthamoeba* actophorin (ADF/Cofilin) with actin filaments. *J. Biol. Chem.* **274,** 15,538–15,546.
13. Kelleher, J. F., Mullins, R. D., and Pollard, T. D. (1998) Purification and assay of the Arp2/3 complex from *Acanthamoeba castellanii*. *Methods Enzymol.* **298,** 42–51.
14. Dayel, M. J., Holleran, E. A., and Mullins, R. D. (2001) Arp2/3 complex requires hydrolyzable ATP for nucleation of new actin filaments. *Proc. Natl. Acad. Sci. USA* **98,** 14,871–14,876.
15. Mullins, R. D., Heuser, J. A., and Pollard, T. D. (1998) The interaction of Arp2/3 complex with actin: nucleation, high affinity pointed end capping, and formation of branching networks of filaments. *Proc. Natl. Acad. Sci. USA* **95,** 6181–6186.
16. Neff, R. J. (1958) Mechanisms of purifying amoebae by migration on agar surfaces. *J. Protozool.* **5,** 226–231.

17. Schuster, F. L. (2002) Cultivation of pathogenic and opportunistic free-living amebas. *Clin. Microbiol. Rev.* **15,** 342–354.
18. Higgs, H. N., Blanchoin, L., and Pollard, T. D. (1999) Influence of the C terminus of Wiskott-Aldrich syndrome protein (WASp) and the Arp2/3 complex on actin polymerization. *Biochemistry* **38,** 15,212–15,222.
19. Zigmond, S. H. (2000) How WASP regulates actin polymerization. *J. Cell Biol.* **150,** F117–F120.
20. Selden, L. A., Kinosian, H. J., Estes, J. E., and Gershman, L. C. (2000) Cross-linked dimers with nucleating activity in actin prepared from muscle acetone powder. *Biochemistry* **39,** 64–74.

Index